AF361735

MOCAST 2021

MOCAST 2021

Editors

Spyridon Nikolaidis
Rodrigo Picos

MDPI • Basel • Beijing • Wuhan • Barcelona • Belgrade • Manchester • Tokyo • Cluj • Tianjin

Editors
Spyridon Nikolaidis
Physics Department
Aristotle University of
Thessaloniki
Thessaloniki
Greece

Rodrigo Picos
Industrial Engineering and
Construction Department
University of Balearic Islands
Palma
Spain

Editorial Office
MDPI
St. Alban-Anlage 66
4052 Basel, Switzerland

This is a reprint of articles from the Special Issue published online in the open access journal *Technologies* (ISSN 2227-7080) (available at: www.mdpi.com/journal/technologies/special_issues/ MOCAST_Technologies).

For citation purposes, cite each article independently as indicated on the article page online and as indicated below:

LastName, A.A.; LastName, B.B.; LastName, C.C. Article Title. *Journal Name* **Year**, *Volume Number*, Page Range.

ISBN 978-3-0365-6294-0 (Hbk)
ISBN 978-3-0365-6293-3 (PDF)

Contents

technologies

Editorial

MOCAST 2021

Spyridon Nikolaidis [1,*] and **Rodrigo Picos** [2,*]

1   Physics Department, Aristotle University of Thessaloniki, 54636 Thessaloniki, Greece
2   Industrial Engineering and Construction Department, University of Balearic Islands, 07122 Palma, Spain
*   Correspondence: snikolaid@physics.auth.gr (S.N.); rodrigo.picos@uib.es (R.P.)

Citation: Nikolaidis, S.; Picos, R. MOCAST 2021. *Technologies* **2022**, *10*, 87. https://doi.org/10.3390/technologies10040087

Received: 13 July 2022
Accepted: 18 July 2022
Published: 20 July 2022

Publisher's Note: MDPI stays neutral with regard to jurisdictional claims in published maps and institutional affiliations.

The International Conference on Modern Circuits and Systems Technologies (MOCAST) was first launched in 2012 inside the framework of a European Project (JEWEL). Its aim was to provide a forum for researchers in Mediterranean and European areas to share and discuss their latest developments. Over the years, MOCAST has consolidated its position as a leading conference, successfully passing even the test of the COVID-19 pandemics in 2021. This special issue is, thus, very special, since it corresponds not only to a recompilation of some of the best papers at the Conference but is also a tribute to all those researchers that kept on working, many times at home with reduced resources, and holding on to the very basic idea of science, which is sharing knowledge.

In this special issue we have twelve papers, covering a very wide range of topics, as in the Conference itself.

In [1], the authors present a circuit able to emulate a memristive system using switched capacitors as the variable resistance and implementing the model of the memristor using stochastic computing.

The second paper [2] explores how 3D integration affects the parasitic coupling using a two-layer 3D pixel as the case of study. Specifically, they use TCAD simulations to study a Back-Side Illuminated, 4T–APS, 3D Sequential Integration pixel with both its photodiode and Transfer Gate at the bottom tier and the other parts of the circuit on the top tier.

The third paper [3] deals with another hot topic, applying AI techniques to help detect the effects of COVID-19 on patients by analyzing X-ray chest images. They propose a robust, lightweight network where the excellent classification of four classes (COVID-19, normal, viral pneumonia, and lung opacity) is performed. The experimental results show that the proposed modified architecture of the model achieves very high classification performance in terms of 21,165 chest X-ray images, and at the same time, it meets real-time constraints in a low-power embedded system.

In [4], the authors present a novel design of current pre-amplifiers suited for visible light communications. In this paper, three pre-amplification topologies for the VLC receiver AFE are presented and compared. All three use bipolar transistors (BJT): the first consists of a single BJT, the second of a double BJT in cascade connection, and the third of a double BJT in Darlington-like connection. In order to validate the performance characteristics of the three topologies, simulation results are provided with respect to the light illumination intensity, the data transmission frequency, and the power consumption.

In [5], Markov chains are used to study the performance of 2D and 3D networks on chips. In this work, a formal way of describing a bufferless NoC topology as a set of discrete-time Markov chains is presented. It is demonstrated that by combining this description with the network average distance, it is possible to obtain the expectation of the number of hops between any pair of nodes in the network as a function of the flit deflection probability. Comparisons between the proposed model and cycle-accurate simulation demonstrate the accuracy achieved by the model, with negligible computational cost. The useful range of the proposed model is quantified, demonstrating that it has an error of less than 10% for a significant proportion (between 33% and 75%) of the injection rate range below saturation. Finally, a simple equation for comparing mesh topologies with a "back-of-the-envelope" calculation is introduced.

In [6], the authors present an IoT system suitable for monitoring and controlling a hydroponic farm. The system is based on three types of sensor nodes. The main (master) node is responsible for controlling the pump, monitoring the quality of the water in the greenhouse, and aggregating and transmitting the data from the slave nodes. Environment-sensing slave nodes monitor the ambient conditions in the greenhouse and transmit the data to the main node. Security nodes monitor activity (movement in the area). The system monitors water quality and greenhouse temperature and humidity, ensuring that crops grow under optimal conditions according to hydroponics guidelines. Remote monitoring for the greenhouse keepers is facilitated by monitoring these parameters via connecting to a website. An innovative fuzzy inference engine determines the plant irrigation duration. The system is optimized for low power consumption in order to facilitate off-grid operation.

Paper [7] deals with the effect of time and amplitude variations on the soft error rate (SER) in VLSI circuits. An accurate SER evaluation is provided based on a SPICE-oriented electrical masking analysis combined with a TCAD characterization process. Furthermore, the proposed work analyzes the effect of a Static Timing Analysis (STA) methodology and the actual interconnection delay on SER evaluation. An analysis of the generated Single Event Multiple Transients (SEMTs) and the circuit operating frequency that are related to the SER estimation is also discussed. Various benchmarks synthesized utilizing 45 nm and 15 nm technology are employed, and the experimental results demonstrate the SER variation as the device node scales down.

Paper [8] proposes a novel schema to implement FIR filters using stochastic computing, showing how this technique is effective in reducing the number of required components, and also the power needed to perform the calculations. Simulation in the spectral domain demonstrates the filter's proper operation and its roll-off behavior, as well as the signal-to-noise ratio improvement using the sigma-delta modulator compared to typical stochastic computing filter realizations. The proposed architecture's hardware advantages are showcased with synthesis results for two FIR filters using FPGA and synopsys tools, while comparisons with standard stochastic computing-based hardware realizations, as well as with conventional binary ones, demonstrate its efficacy.

Paper [9] presents a model for a Tantalum oxide memristor, including parameter estimation, and shows how to apply it to a crossbar. The proposed model is applied and analyzed in hybrid and passive memory crossbars in an LTSPICE environment and is based on the standard Ta_2O_5 memristor model proposed by Hewlett–Packard. The optimal values of coefficients of the tantalum oxide memristor model are derived through the comparison of experimental current–voltage relationships and by using a procedure for parameter estimation. A simplified LTSPICE library model corresponding to the analyzed tantalum oxide memristor is created in accordance with the considered mathematical model. The improved and altered Ta_2O_5 memristor model is tested and simulated in hybrid and passive memory crossbars for a state near a hard-switching operation.

Paper [10] presents an AI-based system able to recognize the art style of a work. The paper presents two different Deep Learning architectures—Vision Transformer and MLP Mixer (Multi-layer Perceptron Mixer)—trained from scratch in the task of artwork style recognition, achieving over 39% prediction accuracy for 21 style classes on the WikiArt paintings dataset. In addition, a comparative study between the most common optimizers was conducted, obtaining useful information for future studies.

Paper [11] proposes a new methodology to encode quantum information into the energy states of a physical system, thus paving the way for the actual implementation of quantum computers. The scheme is based on the notion of encoding logical quantum states using the charge degree of the freedom of the discrete energy spectrum that is formed by introducing impurities in a semiconductor material. They propose a mechanism for performing single-qubit operations and controlled two-qubit operations, providing a mechanism for achieving these operations using appropriate pulses generated by Rabi oscillations. The above architecture is simulated using the Armonk single-qubit quantum computer of IBM to encode two logical quantum states into the energy states of Armonk's qubit and using custom pulses to perform one- and two-qubit quantum operations.

Finally, paper [12] proposes a Lagrangian relaxation method to optimize gate design (sizing and threshold) in a VLSI system, focusing on timing closure. To this end, they transform a Lagrangian-relaxation-based optimizer into a practical incremental timing optimizer that corrects small timing violations with fast runtime without increasing the area/power of the design. The proposed approach is applied to already-optimized designs of the ISPD 2013 benchmarks assuming they experience new timing violations due to local wire rerouting. Experimental results show that, in single corner designs, timing is improved by more than 36% on average, using 45% less runtime. Correspondingly, in a multi-corner context, timing is improved by 39% when compared to the fully-fledged version of the timing optimizer.

We would also like not to miss the opportunity to thank all the attendants for sharing their work, as well as for their passion for research, which makes it worthwhile organizing a Conference.

Conflicts of Interest: The authors declare no conflict of interest.

References

1. De Benito, C.; Camps, O.; Al Chawa, M.; Stavrinides, S.; Picos, R. A switched capacitor memristor emulator using stochastic computing. *Technologies* **2022**, *10*, 39. [CrossRef]
2. Sideris, P.; Peizerat, A.; Batude, P.; Sicard, G.; Theodorou, C. Parasitic coupling in 3D sequential integration: The example of a two-layer 3D pixel. *Technologies* **2022**, *10*, 38. [CrossRef]
3. Sanida, T.; Sideris, A.; Tsiktsiris, D.; Dasygenis, M. Lightweight neural network for COVID-19 detection from chest X-ray images implemented on an embedded system. *Technologies* **2022**, *10*, 37. [CrossRef]
4. Poulis, S.; Papatheodorou, G.; Papaioannou, C.; Sfikas, Y.; Plissiti, M.; Efthymiou, A.; Liaperdos, J.; Tsiatouhas, Y. Effective current pre-amplifiers for visible light communication (VLC) receivers. *Technologies* **2022**, *10*, 36. [CrossRef]
5. Tatas, K. Performance analysis of 2D and 3D bufferless NoCs using Markov chain models. *Technologies* **2022**, *10*, 27. [CrossRef]
6. Tatas, K.; Al-Zoubi, A.; Christofides, N.; Zannettis, C.; Chrysostomou, M.; Panteli, S.; Antoniou, A. Reliable IoT-based monitoring and control of hydroponic systems. *Technologies* **2022**, *10*, 26. [CrossRef]
7. Tsoumanis, P.; Paliaroutis, G.; Evmorfopoulos, N.; Stamoulis, G. Analysis of the impact of electrical and timing masking on soft error rate estimation in VLSI circuits. *Technologies* **2022**, *10*, 23. [CrossRef]
8. Temenos, N.; Vlachos, A.; Sotiriadis, P. Efficient stochastic computing FIR filtering using sigma-delta modulated signals. *Technologies* **2022**, *10*, 14. [CrossRef]
9. Mladenov, V.; Kirilov, S. A simplified tantalum oxide memristor model, parameters estimation and application in memory crossbars. *Technologies* **2022**, *10*, 6. [CrossRef]
10. Iliadis, L.; Nikolaidis, S.; Sarigiannidis, P.; Wan, S.; Goudos, S. Artwork style recognition using vision transformers and MLP mixer. *Technologies* **2022**, *10*, 2. [CrossRef]
11. Ntalaperas, D.; Konofaos, N. Encoding Two-qubit logical states and quantum operations using the energy states of a physical system. *Technologies* **2022**, *10*, 1. [CrossRef]
12. Mangiras, D.; Dimitrakopoulos, G. Incremental Lagrangian relaxation based discrete gate sizing and threshold voltage assignment. *Technologies* **2021**, *9*, 92. [CrossRef]

technologies

Article

A Switched Capacitor Memristor Emulator Using Stochastic Computing [†]

Carola de Benito [1,2], Oscar Camps [1], Mohamad Moner Al Chawa [3], Stavros G. Stavrinides [4] and Rodrigo Picos [1,2,*]

[1] Industrial Engineering and Construction Department, Balearic Islands University, 07122 Palma, Spain; carol.debenito@uib.es (C.d.B.); oscar.camps@uib.es (O.C.)
[2] Health Institute of the Balearic Islands, 07121 Palma, Spain
[3] Institute of Circuits and Systems, Technical University of Dresden, 01062 Dresden, Germany; mohamad_moner.al_chawa@tu-dresden.de
[4] School of Science and Technology, International Hellenic University, 57001 Thessaloniki, Greece; s.stavrinides@ihu.edu.gr
* Correspondence: rodrigo.picos@uib.es
† This paper is an extended version of our paper published in de Benito, C.; Camps, O.; Al Chawa, M.M.; Stavrinides, S.G.; Picos, R. A Stochastic Switched Capacitor Memristor Emulator. 2021 10th International Conference on Modern Circuits and Systems Technologies (MOCAST). IEEE, 2021.

Abstract: Due to the increased use of memristors and their many applications, the use of emulators has grown in parallel to avoid some of the difficulties presented by real devices, such as variability and reliability. In this paper, we present a memristive emulator designed using a switched capacitor (SC), that is, an analog component/block and a control part or block implemented using stochastic computing (SCo) and therefore fully digital. Our design is thus a mixed signal circuit. Memristor equations are implemented using stochastic computing to generate the control signals necessary to work with the controllable resistor implemented as a switched capacitor.

Keywords: memristor; emulator; analog design; switched capacitor; stochastic computing; mixed signal

Citation: de Benito, C.; Camps, O.; Al Chawa, M.M.; Stavrinides, S.G.; Picos, R. A Switched Capacitor Memristor Emulator Using Stochastic Computing. *Technologies* **2022**, *10*, 39. https://doi.org/10.3390/technologies10020039

Academic Editors: Roberto Osellame and Petra Paiè

Received: 1 December 2021
Accepted: 15 February 2022
Published: 2 March 2022

Publisher's Note: MDPI stays neutral with regard to jurisdictional claims in published maps and institutional affiliations.

1. Introduction

Leon Chua defined the memristor theoretically in 1971 [1]. The term "memristor" is constructed from the words memory and resistor and completes the relationships provided by the capacitor, inductor and resistor between current, voltage, flux and charge. Leon Chua introduces, therefore, this fourth passive component to complete the set, proposing that the memristor is defined by a nonlinear relationship between charge and flux. However, it was not until 2008 that it could be implemented [2]. Since then, its use has been increasing and its fields of application have also increased. It is a promising but very recent device, which implies that there are many studies that must be carried out to understand well the operating mechanisms and develop new technologies to avoid some of the problems presented, such as the variability and life time.

It is because of the problems that real devices present that the development of memristive emulators is booming.

Emulators reproduce the operating characteristics of the memristor by eliminating the aforementioned problems, therefore allowing for the development of more complex and reliable systems [3]. The memristor behavior which is imitated can be an ideal memristor or actual device, depending on the implementation. If we are focused on their field of application, emulators have different characteristics, although there are two main lines of study: analog emulators and digital emulators.

Many works develop fully analog emulators; for example, in [4], a memristive system was implemented and results demonstrated that it was very easy to fabricate in academic laboratories through classical electrical components from circuit theory. In [5], an emulator is implemented with transistors, resistors and diodes, and it operates in passive mode. Other examples like [6,7] use amplifiers in their models. In general, analog systems need more power consumption. The volatility of the system is worse than in the digital case, but they present a good implementation of the variable resistance with which the emulator memristance is described.

On the other hand, digital memristor systems emulators can be implemented in FPGAs (or ASICS) [8–10]. Their main advantages are that they present short simulation times and better control of the behaviour of the emulator. However, their variable resistance implementation poses a problem. In digital emulators, it is much easier to define the model, but precision is lost (limited number of bits), and there is usually a need for more computational power than the analog equivalent. For a review of different state-of-the-art emulators, the interested reader can see, for instance, [3] or [11].

For many years, the scientific community has been greatly interested in developing different computer architectures. It seeks, among other things, to change the structure of serial calculations and perform operations in parallel. A high degree of parallelism allows for faster execution using less complex elements or using approximations even if precision is lost. Due to the large number of data and operations that must be carried out by current computer systems, the well-known Von Neumann architecture [12] is not a good choice to use due to its high consumption of time of computation and energy.

One of the alternatives, as mentioned above, is to use non-deterministic computing methods, including stochastic computing (SCo). Its main differentiating trait is that it uses random variables to represent quantities. Probabilistic logic was introduced by Von Neumann in 1956 [13], expanding the previous work of R.S. Pierce in 1952 [14]. However, it was not until the 1960s that progress in electronics and computing allowed for its actual implementation [15,16]. Nowadays, there are many proposals in the literature using this approach in different fields: image processing [17–19], data compression [20], arithmetic calculations [21,22], control [23], and A/D conversion [24], to mention just a few.

In this framework, the representation of data is performed in a probabilistic way using Boolean quantities that are switched in random way during a time. Numbers are represented as random (0,1) vectors. The average value of these vectors is correlated to the number represented [25]. These vectors are referred to as stochastic logic number (SLN). This representation makes it possible to reduce the area occupied, since complex functions occupying a large space such as multiplication can be reduced to a single logic gate with great savings in terms of power and area [26]. To create the SLN, a random number generator (RNG) is needed, and for the designers, it is a challenge to use the lowest number of these so as to not increase the area of the chip. The number of RNGs is related to the fact that the operations in SCo are different depending on the encoding of the number and if the signals are correlated or not (statistical dependent or not). To guarantee uncorrelated SLNs, different RNGs must be employed to generate each stochastic signal.

In this work, we design, simulate and implement a mixed-signal memristor emulator, improving the version presented in [27]. Specifically, in this paper, we improve the theoretical discussion, including the description of the stochastic blocks, and we also present some experimental results. The proposed emulator consists of two blocks, taking advantage of the best features of each design part. In the analog block, a switched capacitor is used to implement a variable resistor, and in the digital one, that is, the control block, we use stochastic computation. The simulation is done with Matlab to implement the functionality of both the analog block, similar to that used in [28], and of the control block. For the experimental implementation, a quadruple analog switch HCF4066FE and a DE0-Nano FPGA have been used.

This paper is organized as follows: the next section describes the generalities of memristors, memristor emulators and stochastic logic operations; in the third section, the model is developed and simulated; the fourth section deals with the experimental implementation; and, finally, the last section discusses the work.

2. Theoretical Background

2.1. Memristor Mathematical Description

A memristor is a two-terminal device whose resistance (conductance) can change its value when a voltage or current signal is applied. In addition, the value of the resistance (conductance) of the device also depends on its past history and is named memristance (M) (memconductance (G)). The concept of the memristor was extended by Chua in 1976 to memristive systems to explain the behavior of observed systems [29], for instance, in nature. Nowadays the classification of memristors includes ideal, generic and extended memristor [30].

The most general class is the extended memristor, which includes the others. The dynamic of this class is described using internal variables that determine the internal state of the memristor; these variables, can be for example, temperature or geometrical parameters, depending on the system. The memristor can be voltage- or current-controlled, depending on the input source. On the other hand, in [31], Corinto et al. proposed a mathematical description in the charge flux domain instead of the voltage and current domain. We use for our emulator the equations describing a voltage-controlled extended memristor in the charge flux domain. These are:

$$i = G(\phi, v, \mathbf{x}) \cdot v \tag{1}$$

$$\frac{d\mathbf{x}}{dt} = \mathbf{g}(\phi, v, \mathbf{x}) \tag{2}$$

$$\frac{d\phi}{dt} = v \tag{3}$$

The memconductance (G), which can be nonlinear, is the inverse of the memristance of the device M, v is the voltage between its terminals, i is the current, ϕ is the flux (i.e., the first momentum of voltage), and $\mathbf{x}$ represents other possible state variables.

Finally, it is also important to mention that the memristors present some characteristic fingerprints distinguishing those of other dynamic systems [31,32]:

1. As Leon Chua noted in [33]: "If it's NOT pinched, it's NOT a memristor". The *i-v* curve obtained when a periodic signal with zero DC component (voltage or current) is applied to the memristor shows a pinched (at the (v = 0, i = 0) point) hysteresis loop;
2. The area of the hysteresis loop should tend to zero for higher frequencies, as noted in [31]. The behavior at low frequencies depends on the specifics of the memristor, and there may even exist a frequency where the loop area is maximum [3].

On the other hand, the emulator function must be to mimic the memristor behavior; this is to show it fingerprints. The emulator can be implemented in analog, digital or mixed formats. It is crucial that the circuit implements, among others, the internal state variables, (vector $\mathbf{x}$) in Equations (1) and (2). These internal variables must be included as electrical variables in the emulator and are assumed to be isolated from a direct interaction with the outside. They are used, together with the electrical variables (i.e. voltage and flux), to calculate the value of the equivalent memconductance (G) or memristance (M).

Notice that we have implemented the ideal definition of a memristor, with a simple relationship between memristance and flux. Other models, even those oriented to the simulation of actual physical systems as, for instance, in [34–36], could also be implemented. The main difference of this case with the one presented here would be the implementation of the non-basic mathematical operations. This could be done using, for instance, the different circuits proposed in [37–39] for division and the associated square root calculation, or in [40] for arbitrary function approximation.

2.2. Stochastic Logic Operations

There are four methods that are used to encode numbers in SCo: unsigned classical stochastic encoding (UCSE), signed classical stochastic encoding (SCSE), unsigned extended stochastic encoding (UESE) and signed extended stochastic encoding (SESE) [41,42]. Implementation of different functions strongly depends on the chosen encoding. As an example, when using UCSE, an AND gate is used to implement the product of two inputs, while a XNOR gate performs that operation when using the SCSE encoding, as shown in Figure 1. Moreover, the same logical gate can perform different operations depending on whether the random signals generated are correlated or not. As an example, with USCE encoding, a two-inputs AND gate is used to implement a multiplication for uncorrelated inputs, while it provides the minimum value of the two inputs if they are correlated. In the present work, we will be using SCSE, and, thus, our numbers will lay in the real $(-1..1)$ domain. Thus, multiplication requires the use of an XNOR gate.

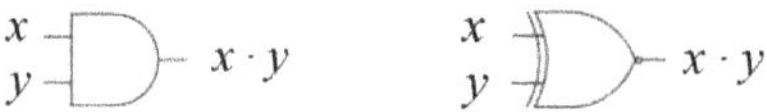

Figure 1. Basic implementation scheme of a SC multiplier in the $(0..1)$ range (AND gate, **left**) and in the $(-1..1)$ range (XNOR gate, **right**).

Performing an addition is slightly more complex, because numbers with a probability higher than one cannot be represented, and we may need to add $1 + 1 = 2$. Thus, it is better to implement an alternate form as $(x + y)/2$, which at most will output a value of 1. This operation is usually implemented using a multiplexer, as shown in Figure 2a, where the p (0.5) means a signal with a probability of 0.5 to be '1' or '0'. This signal can be generated using one of the bits from the RNG, thus needing no additional circuitry. The same gate is used for the $(0..1)$ and the $(-1..1)$ domains.

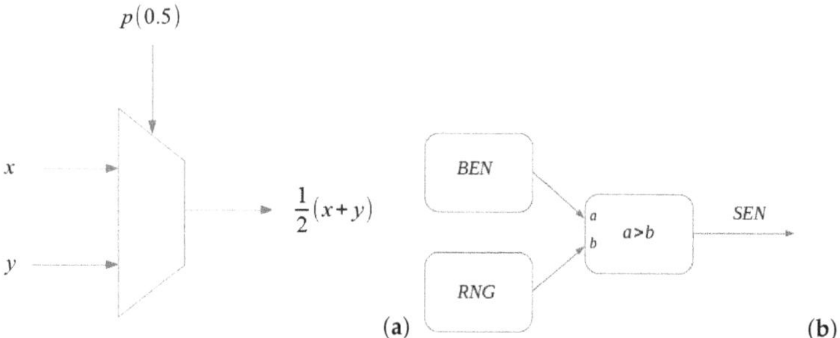

Figure 2. Basic implementation schemes (**a**) of a SC adder using a multiplexer and (**b**) a stochastic number generator (SGN) which converts a binary encoded number (BEN) to a stochastic encoded number (SEN) using a random number generator (RNG).

Other more complex operations (division [23], square roots [23], etc.) may also be found in the literature, but are not presented here for the sake of clarity. Finally, the conversion of a number encoded as a classical number can be translated to a stochastic representation using the schema presented in Figure 2.

3. Memristor Emulator Design

3.1. Theoretical Design

As mentioned above, our system has been implemented in two parts [27,28]. First, we implemented an analog system including the switched capacitor module (SC), as shown in Figure 3, whose equivalent resistance R_{eq} is described by Equation (4). In this case, both control external signals S_1 and S_2 are equal, with a lag of 180 deg. [43]. The second part is a digital module implementing the control part in stochastic logic, as will be discussed below.

$$R_{eq} = \frac{1}{f_S C} \frac{1 + e^{\left(\frac{D\,T}{\tau}\right)}}{1 - e^{\left(\frac{D\,T}{\tau}\right)}} \tag{4}$$

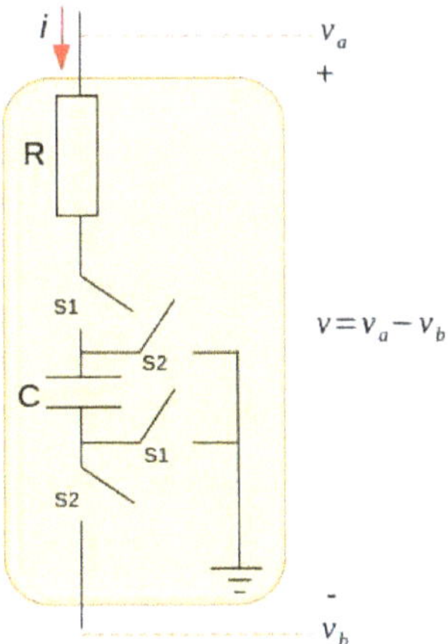

Figure 3. Switched capacitor (SC) circuit schematic.

T and f_s are the period and the frequency of the controlling signal S_1 and S_2, D is the duty cycle with values between 0 and 1, C is the capacitance value and τ is the time constant, that is, $R_{tot}C$ where C is the capacitor and R_{tot} is the total resistance of the circuit taking into account the parasitic ones.

For our design, in the charge flux domain, it is necessary to calculate the flux from the voltage of the terminals of the SC as a first step. Once this is done, then the relationships between flux and charge are used to obtain the duty cycle (D) that varies the equivalent resistance of the SC. The digital block is the responsible for all these steps.

For this purpose, a series of approximations shall be done to Equation (4). The conductance (G) ($G = 1/R_{eq}$) can be rewritten as:

$$G = f_S C \frac{e^{\left(-\frac{x}{2}\right)} - e^{\left(\frac{x}{2}\right)}}{e^{\left(-\frac{x}{2}\right)} + e^{\left(\frac{x}{2}\right)}} = f_S \cdot C \cdot tanh\left(\frac{x}{2}\right) \tag{5}$$

where $x = DT/\tau$.

The use of a first order Taylor expansion of $tanh(x/2)$ allows us to get a simpler expression. For this, it is necessary to take into account that the decay time of the system is much longer than the control signal period. Thus, we can obtain a simpler equation describing the conductance G:

$$G = f_S C \frac{DT}{2\tau} \tag{6}$$

It is important to notice that this last equation implies that conductance is linearly dependent on the duty cycle D.

To calculate the flux, the digital block converts each voltage terminal of the SC (v_a and v_b in Figure 3) to non-correlated random values. Then, the corresponding value $v_a - v_b$ is accumulated into a counter, which acts as the integrator. Notice that since we are using stochastic computing, this up/down counter needs to count only one up ($v_a > v_b$), one down ($v_a < v_b$), or remain the same ($v_a = v_b$). To implement the memristor device, it is necessary to use an equation to describe the relationship between flux and charge. In this work the simplest relation is used:

$$Q = M\phi^2 \tag{7}$$

where M is a constant. This equation does not include any internal variables. Applying the fourth derivative of the equation, the conductance is:

$$i = 2M\phi\frac{d\phi}{dt} = 2M\phi v \implies G = 2M\phi \tag{8}$$

Matching Equations (6) and (8), the relation between the duty cycle and flux is:

$$D = \frac{4M\tau}{f_S CT}\phi = K\phi \tag{9}$$

K is therefore a constant value.

To control the analog block, it is the SC; therefore, the duty cycle (D) must be used. The duty cycle is calculated by the digital block from ϕ according to Equation (9) as a stochastic value. To use it, the average value of D is calculated to determine R_{eq} with Equation (4).

The emulator block design scheme including the two parts of the design, analog and digital, is shown in Figure 4. The part corresponding to the digital block implemented in stochastic computing is shown as a circuit in Figure 5.

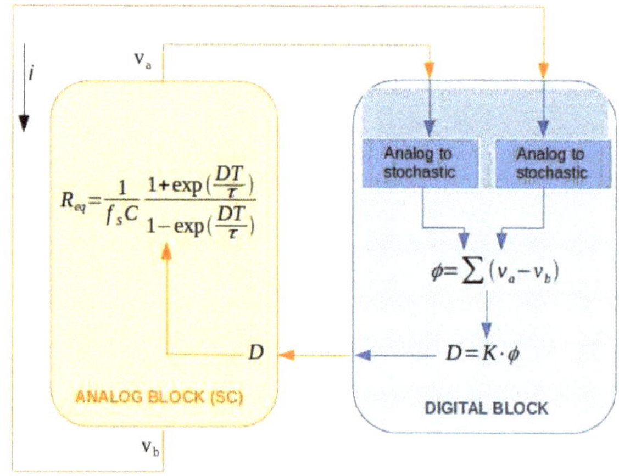

Figure 4. Swiched capacitor memristor emulator (SCME) block diagram.

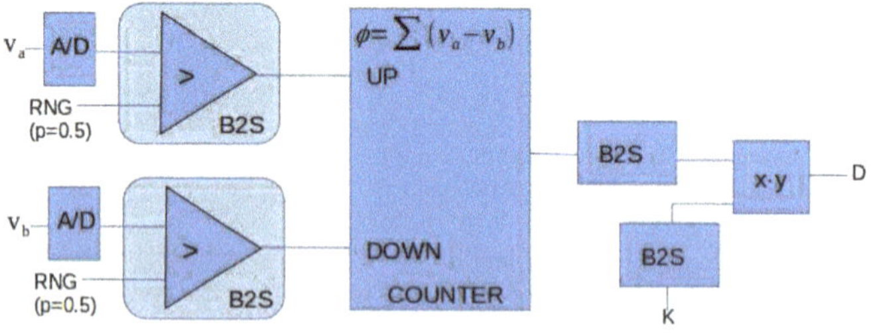

Figure 5. Control block implementation using stochastic computing.

3.2. Simulation Results

In order to be considered as a memristor, the emulator must present two characteristic fingerprints [3,32,33]: (1) a pinched loop (2) whose area changes with frequency.

Figure 6 presents the $i - v$ of the emulator under inputs of different frequency using 16 bits for the stochastic representation. It is apparent from this figure that the curves are pinched at the origin and that the loop area changes with frequency. Thus, we can consider that the two fingerprints are present.

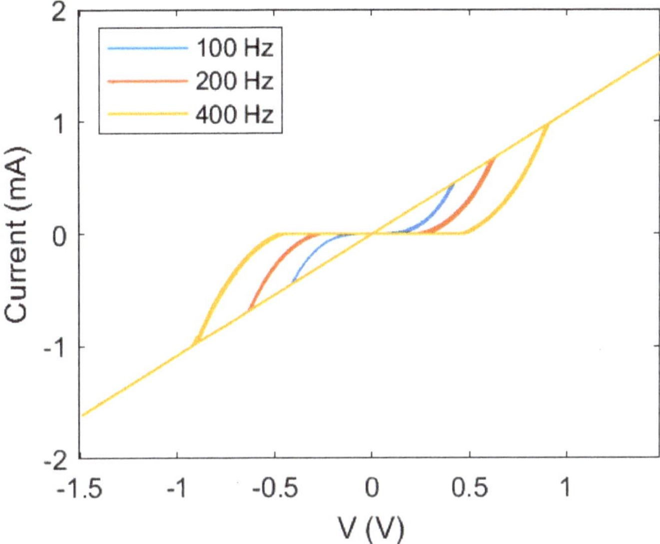

Figure 6. Simulated $i - v$ characteristic curves of the memristor implemented using Figure 4. Three different frequencies are shown in different colors.

Because of the way it is constructed, the emulator reaches a saturation for the conductance. This is due to the maximum value of $D = 1$, and can be clearly seen at low frequencies, where the maximum value of flux is reached faster. This may also be seen in Figure 7, where the behavior of the Q versus ϕ near the origin is quadratic, as can be expected from (7), but it is also seen that its behavior changes to linear after a maximum value for $D = 1$ is reached.

It has to be noted that there is a small noise present caused by the stochastic nature of the system, as discussed above. This noise nearly disappears in the saturation, since the counter is practically constant, even though a small ripple is present caused by the stochastic internal behavior. This noise is greatly reduced in the charge and flux domain (Figure 7, because of the integration.

Finally, the current signal for different frequencies is shown in Figure 8. As can be seen there, the maximum conductance (related to the maximum value of the current) is lower for higher values of frequency, as expected.

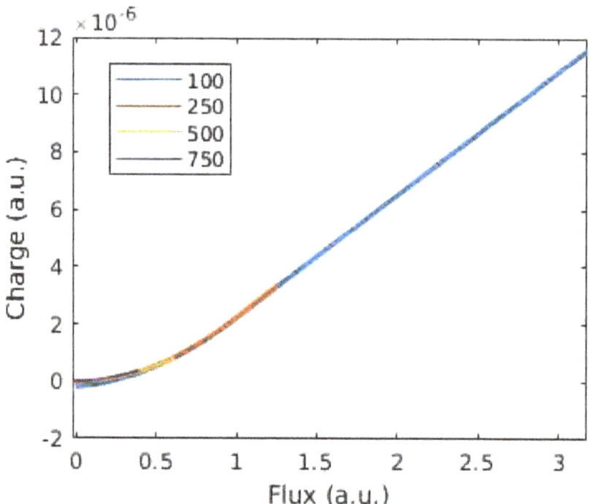

Figure 7. $Q - \phi$ characteristics of the memristor implemented using Figure 4. The different frequencies (in arbitrary units) are shown in different colors.

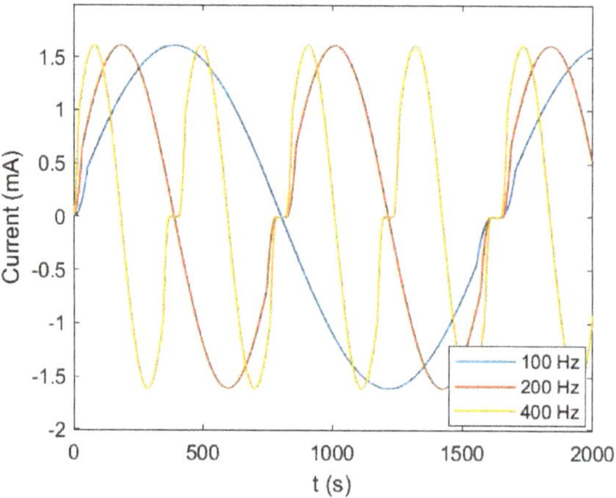

Figure 8. Current signal (response) for different frequencies, as obtained from the simulation. The three different frequencies are shown in different colors and correspond to the ones shown in Figure 6.

4. Experimental Implementation

4.1. Experimental Setup

In order to test the proposed circuit, we have implemented a setup similar to that of [28]. We have used a quadruple analog switch HCF4066FE driven by a DE0-Nano FPGA. The analog switch has a working voltage between 0.5 V to 22 V and can switch at a maximum frequency of 25 kHz when the power supply is 3.3 V. The FPGA generated the control signals S_1 and S_2 using 2 of its 3.3 V digital output pins. Additionally, we have also used a 1 kΩ shunt resistor, along with a 15 µF capacitor. The implemented circuit is shown in Figure 9. The conversion from analog to stochastic was performed by first converting from analog to digital using two of the on-board available A/D and then converting this digital value into stochastic, as described above. We have used 16 bits for the stochastic representation, and the needed random numbers were created using a public implementation of the Mersenne twister algorithm [44].

Figure 9. Physical implementation of the circuit on a prototyping board.

An AFG320 arbitrary signal generator was used to generate the input signal, while two oscilloscopes were used to monitor the full system. An oscilloscope monitored the control signals of the HCF4066FE, while the other oscilloscope was used to monitor the voltage through the shunt resistance of 1 kΩ to obtain the current and also the input voltage, defined as the difference between the two input terminals.

4.2. Experimental Results

The system has been tested using different input frequencies: 100, 200 and 400 Hz. The internal behaviour of the circuit is depicted in Figures 10 and 11, which depict, respectively, the control signals S_1 and S_2 in one of these cases and the waveform corresponding to the three least significant bits of the counter.

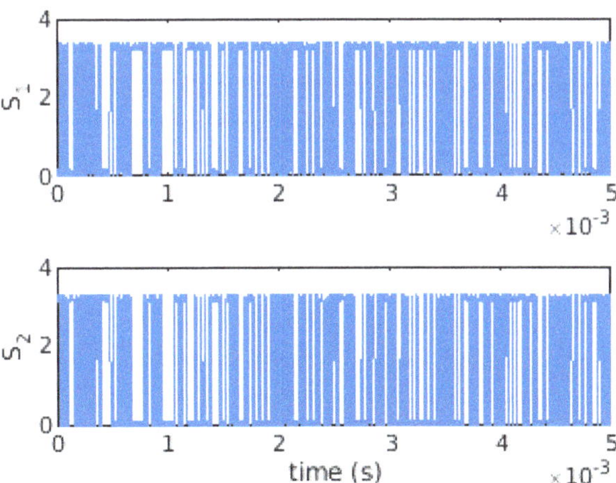

Figure 10. Stochastic signals S_1 and S_2 generated by the control circuit.

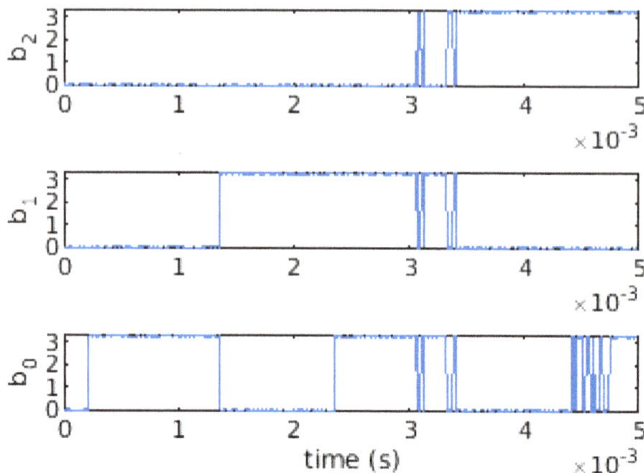

Figure 11. Three least significant bits of the counter (b_0 is the least significant bit) at a specific time.

The temporal behavior of the current in these three cases is shown in Figure 12. The currents are clearly nonlinear because of memory: if they were nonlinear due to other effects, then they would be symmetrical, which they are not. In addition, they are showing a dependence on the frequency, as expected for a memristor.

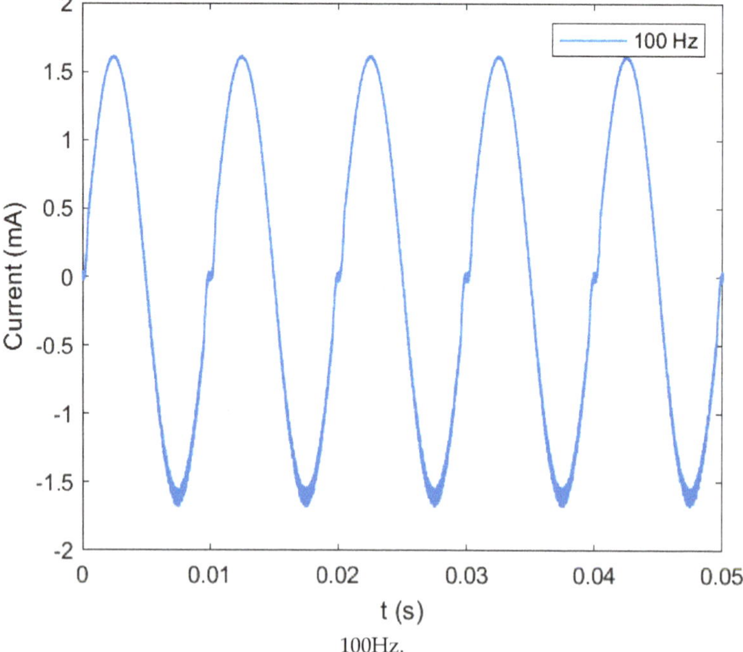

Figure 12. *Cont.*

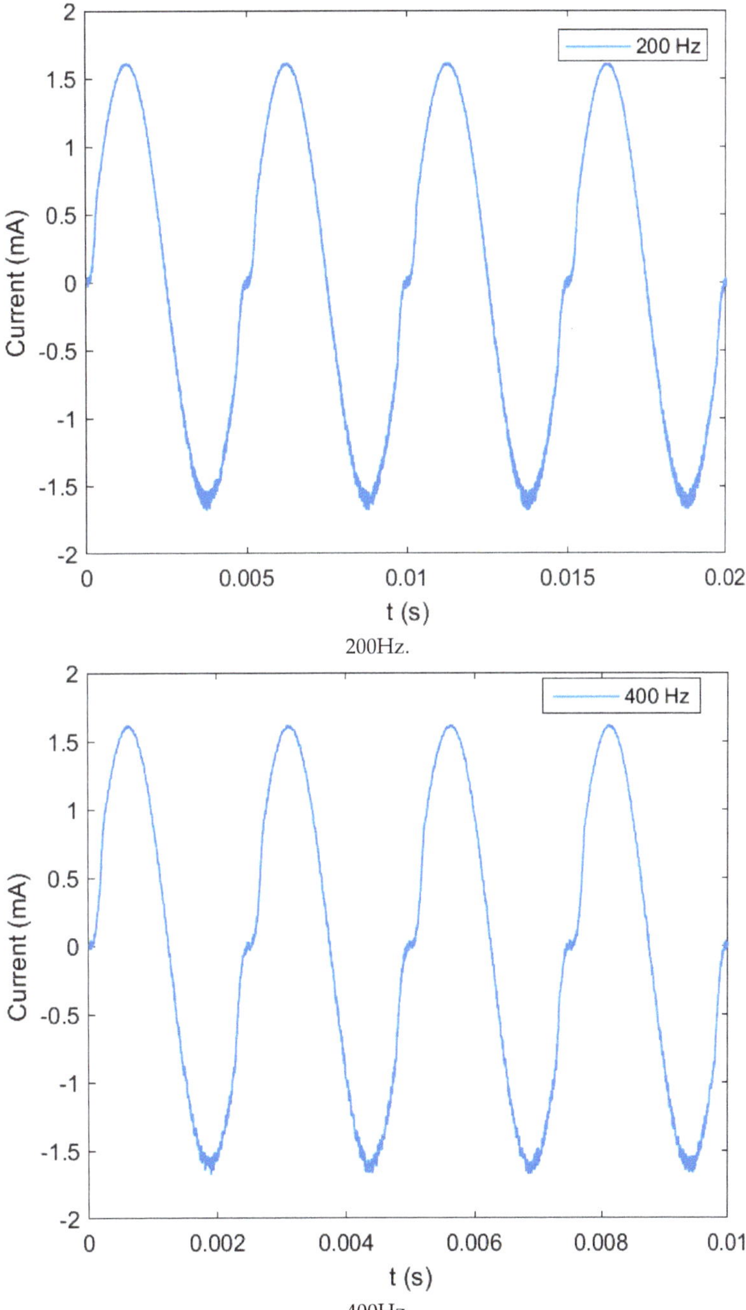

200Hz.

400Hz.

Figure 12. Temporal graphs of the measured response (current signals) of the realized memristor at 3 different frequencies corresponding to the simulated frequencies for driving sine voltage of (**a**) 100 Hz, (**b**) 200 Hz, and (**c**) 400 Hz.

The experimental I-V loops are depicted in Figure 13. On the left figure, the three experimental I-V curves for the corresponding frequencies in Figure 6 (simulations) appear, while the right picture shows an oscilloscope snapshot in the typical case of a 400 Hz driving sine voltage. It is apparent that, in all cases, the experimental fingerprint of a memristor, i.e., the pinched loop [33], is clearly demonstrated. This means that the device has a resistive behavior (it is pinched, which means no current when no voltage is applied), and that this resistance has a memory effect (there is a loop, which means that there are two possible values of the resistance and, hence, the current, for each voltage input value).

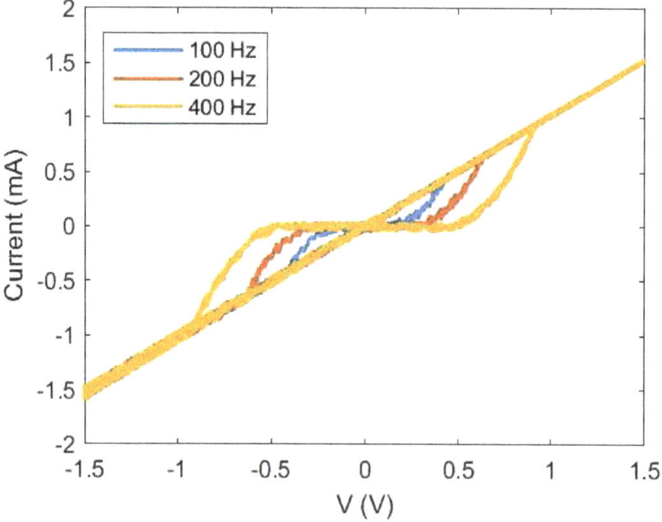

Figure 13. Measured I-V signals at different frequencies.

Finally, it has to be noted that the area of the loop changes with frequency, with the higher area corresponding to higher frequencies. This is caused by the saturation of the internal counter that corresponds to the flux integral (Equation (3) and Figure 5), which leads to a linear behaviour once the maximum value is reached.

5. Discussion

As discussed above, the design and implementation of memristor emulators is an active research field. In this paper, we have made a contribution to this area by presenting the design, simulation, and experimental implementation of such an emulator. Our proposal is based on using switched capacitors to implement the variable resistor and on using stochastic computing to implement the control part.

The switched capacitor block has been implemented using standard off-the-shelf components with a maximum switching frequency of 25 kHz. The control signals at this frequency are generated inside the control block, which has been implemented into a DE0-nano FPGA. The FPGA reads the analog inputs (the input voltage of the analog block) using its built-in AD converters.

As a first step, we have shown using MATLAB simulation that the design is sound and can implement a system showing the expected fingerprints of a memristor: a closed loop, pinched at the origin. Finally, we have experimentally implemented the design. This actual implementation has been tested using sinusoidal waveforms of different frequencies, and it has behaved as expected. The system shows the memristor fingerprints with noise induced by the switching, as expected.

Thus, the proposed emulator has been shown to perform with its expected behavior, being a promising alternative to be implemented as an IP block into IC designs, since it is a very simple design requiring a lower number of digital gates than similar designs using conventional arithmetic implementations. This implementation would allow the increase of the limiting factor of the switching frequency at 25 kHz caused by the use of a discrete component, and would also proportionally increase the working frequency of the emulator.

Author Contributions: Conceptualization, C.d.B. and R.P.; methodology, C.d.B. and R.P.; software, C.d.B., S.G.S. and R.P.; validation, C.d.B., M.M.A.C. and R.P.; formal analysis, C.d.B. and R.P.; investigation, C.d.B., O.C., M.M.A.C. and R.P.; resources, C.d.B. and R.P.; data curation, C.d.B. and R.P.; writing—original draft preparation, C.d.B., O.C., M.M.A.C., S.G.S. and R.P.; writing—review and editing, C.d.B., O.C., M.M.A.C., S.G.S. and R.P.; visualization, C.d.B., O.C., M.M.A.C. and R.P.; supervision, C.d.B., S.G.S. and R.P.; project administration, C.d.B. and R.P.; funding acquisition, C.d.B. and R.P. All authors have read and agreed to the published version of the manuscript.

Funding: Some of the authors wish to acknowledge support from the DPI2017-86610-P and TEC2017-84877-R projects awarded by the MICINN, as well as the partial support by the FEDER program.

Institutional Review Board Statement: Not applicable.

Informed Consent Statement: Not applicable.

Conflicts of Interest: The authors declare no conflict of interest. The funders had no role in the design of the study; in the collection, analyses, or interpretation of data; in the writing of the manuscript, or in the decision to publish the results.

References

1. Chua, L. Memristor-the missing circuit element. *IEEE Trans. Circ. Theory* **1971**, *18*, 507–519. [CrossRef]
2. Strukov, D.B.; Snider, G.S.; Stewart, D.R.; Williams, R.S. The missing memristor found. *Nature* **2008**, *453*, 80–83. [CrossRef] [PubMed]
3. Stavrinides, S.G.; Picos, R.; Corinto, F.; Al Chawa, M.M.; de Benito, C. Implementing memristor emulators in hardware. In *Mem-Elements for Neuromorphic Circuits with Artificial Intelligence Applications*; Academic Press: Cambridge, MA, USA, 2021; pp. 17–40.
4. Ascoli, A.; Corinto, F.; Tetzlaff, R. A class of versatile circuits, made up of standard electrical components, are memristors. *Int. J. Circuit Theory Appl.* **2016**, *44*, 127–146. [CrossRef]
5. Kalomiros, J.; Stavrinides, S.G.; Corinto, F. A two-transistor non-ideal memristor emulator. In Proceedings of the Modern Circuits and Systems Technologies (MOCAST), 2016 5th International Conference, Thessaloniki, Greece, 12–14 May 2016; pp. 1–4.
6. Kim, H.; Sah, M.P.; Yang, C.; Cho, S.; Chua, L.O. Memristor emulator for memristor circuit applications. *IEEE Trans. Circ. Syst. I Regul. Pap.* **2012**, *59*, 2422–2431.
7. Li, Z.; Zeng, Y.; Ma, M. A novel floating memristor emulator with minimal components. *Act. Passiv. Electron. Components* **2017**, *2017*, 1609787. [CrossRef]
8. Vourkas, I.; Abusleme, A.; Ntinas, V.; Sirakoulis, G.C.; Rubio, A. A Digital Memristor Emulator for FPGA-Based Artificial Neural Networks. In Proceedings of the Verification and Security Workshop (IVSW), Sant Feliu de Guixols, Spain, 4–6 July 2016; pp. 1–4.
9. Ranjan, R.; Ponce, P.M.; Kankuppe, A.; John, B.; Saleh, L.A.; Schroeder, D.; Krautschneider, W.H. Programmable memristor emulator asic for biologically inspired memristive learning. In Proceedings of the Telecommunications and Signal Processing (TSP), 2016 39th International Conference, Vienna, Austria, 27–29 June 2016; pp. 261–264.
10. Kolka, Z.; Vavra, J.; Biolkova, V.; Ascoli, A.; Tetzlaff, R.; Biolek, D. Programmable Emulator of Genuinely Floating Memristive Switching Devices. In Proceedings of the 2019 26th IEEE International Conference on Electronics, Circuits and Systems (ICECS), Genova, Italy, 27–29 November 2019; pp. 217–220.
11. Romero, F.J.; Ohata, A.; Toral-Lopez, A.; Godoy, A.; Morales, D.P.; Rodriguez, N. Memcapacitor and Meminductor Circuit Emulators: A Review. *Electronics* **2021**, *10*, 1225. [CrossRef]
12. Von Neumann, J. First draft of a report on the EDVAC. *IEEE Ann. Hist. Comput.* **1993**, *15*, 27–75. [CrossRef]
13. Von Neumann, J. Probabilistic logics and the synthesis of reliable organisms from unreliable components. *Autom. Stud.* **1956**, *34*, 43–98.
14. Von Neumann, J.; Pierce, R.S. *Lectures on Probabilistic Logics and the Synthesis of Reliable Organisms from Unreliable Components*; California Institute of Technology: Pasadena, CA, USA, 1952.
15. Gaines, B.R. Stochastic computing. In Proceedings of the Spring Joint Computer Conference, New York, NY, USA, 18–20 April 1967; pp. 149–156.
16. Poppelbaum, W.; Afuso, C.; Esch, J. Stochastic computing elements and systems. In Proceedings of the Fall Joint Computer Conference, Anaheim, CA, USA, 14–16 November 1967; pp. 635–644.

17. Fick, D.; Kim, G.; Wang, A.; Blaauw, D.; Sylvester, D. Mixed-signal stochastic computation demonstrated in an image sensor with integrated 2D edge detection and noise filtering. In Proceedings of the IEEE 2014 Custom Integrated Circuits Conference, San Jose, CA, USA, 15–17 September 2014; pp. 1–4.
18. Camps, O.; Stavrinides, S.G.; Picos, R. Efficient Implementation of Memristor Cellular Nonlinear Networks using Stochastic Computing. In Proceedings of the 2020 European Conference on Circuit Theory and Design (ECCTD), Sofia, Bulgaria, 7–10 September 2020; pp. 1–4.
19. Camps, O.; al Chawa, M.M.; Stavrinides, S.G.; Picos, R. Stochastic Computing Emulation of Memristor Cellular Nonlinear Networks. **2021**. Preprints.
20. Wang, R.; Han, J.; Cockburn, B.; Elliott, D. Stochastic circuit design and performance evaluation of vector quantization. In Proceedings of the 2015 IEEE 26th International Conference on Application-Specific Systems, Architectures and Processors (ASAP), Toronto, ON, Canada, 27–29 July 2015; pp. 111–115.
21. Yuan, B.; Wang, Y.; Wang, Z. Area-efficient scaling-free DFT/FFT design using stochastic computing. *IEEE Trans. Circ. Syst. II Express Briefs* **2016**, *63*, 1131–1135. [CrossRef]
22. Camps, O.; Stavrinides, S.G.; Picos, R. Stochastic Computing Implementation of Chaotic Systems. *Mathematics* **2021**, *9*, 375. [CrossRef]
23. Toral, S.; Quero, J.; Franquelo, L.G. Digital stochastic realization of complex analog controllers. *IEEE Trans. Ind. Electron.* **2002**, *49*, 1101–1109.
24. Toral, S.; Quero, J.; Ortega, J.; Franquelo, L. Stochastic A/D sigma-delta converter on FPGA. In Proceedings of the 42nd Midwest Symposium on Circuits and Systems (Cat. No. 99CH36356), Las Cruces, NM, USA, 8–11 August 1999; Volume 1, pp. 35–38.
25. Toral, S.; Quero, J.; Franquelo, L. Stochastic pulse coded arithmetic. In Proceedings of the 2000 IEEE International Symposium on Circuits and Systems (ISCAS), Geneva, Switzerland, 28–31 May 2000; Volume 1, pp. 599–602.
26. Moons, B.; Verhelst, M. Energy-efficiency and accuracy of stochastic computing circuits in emerging technologies. *IEEE J. Emerg. Sel. Top. Circ. Syst.* **2014**, *4*, 475–486. [CrossRef]
27. De Benito, C.; Camps, O.; Al Chawa, M.; Stavrinides, S.; Picos, R. A Stochastic Switched Capacitor Memristor Emulator. In Proceedings of the 2021 10th International Conference on Modern Circuits and Systems Technologies (MOCAST), Thessaloniki, Greece, 5–7 July 2021; pp. 1–4.
28. Svetoslavov, G.; Camps, O.; Stavrinides, S.G.; Picos, R. A Switched Capacitor Memristive Emulator. *IEEE Trans. Circ. Syst. II Express Briefs* **2020**, *68*, 1463–1466. [CrossRef]
29. Chua, L.O.; Kang, S.M. Memristive devices and systems. *Proc. IEEE* **1976**, *64*, 209–223. [CrossRef]
30. Leon, C. Everything you wish to know about memristors but are afraid to ask. *Radioengineering* **2015**, *24*, 319.
31. Corinto, F.; Civalleri, P.P.; Chua, L.O. A theoretical approach to memristor devices. *IEEE J. Emerg. Sel. Top. Circ. Syst.* **2015**, *5*, 123–132. [CrossRef]
32. Biolek, D.; Biolek, Z.; Biolková, V.; Kolka, Z. Some fingerprints of ideal memristors. In Proceedings of the Circuits and Systems (ISCAS), 2013 IEEE International Symposium, Beijing, China, 19–23 May 2013; pp. 201–204.
33. Chua, L. If it's pinched it's a memristor. *Semicond. Sci. Technol.* **2014**, *29*, 104001. [CrossRef]
34. Ielmini, D.; Milo, V. Physics-based modeling approaches of resistive switching devices for memory and in-memory computing applications. *J. Comput. Electron.* **2017**, *16*, 1121–1143. [CrossRef]
35. Chang, K.C.; Chang, T.C.; Tsai, T.M.; Zhang, R.; Hung, Y.C.; Syu, Y.E.; Chang, Y.F.; Chen, M.C.; Chu, T.J.; Chen, H.L.; et al. Physical and chemical mechanisms in oxide-based resistance random access memory. *Nanoscale Res. Lett.* **2015**, *10*, 120. [CrossRef]
36. Williams, R.S.; Pickett, M.D.; Strachan, J.P. Physics-based memristor models. In Proceedings of the Circuits and Systems (ISCAS), 2013 IEEE International Symposium, Beijing, China, 19–23 May 2013; pp. 217–220.
37. Chen, T.H.; Hayes, J.P. Design of Division Circuits for Stochastic Computing. In Proceedings of the 2016 IEEE Computer Society Annual Symposium on VLSI (ISVLSI), Pittsburgh, PA, USA, 11–13 July 2016; pp. 116–121.
38. Mitra, S.; Banerjee, D.; Naskar, M.K. A Low Latency Stochastic Square Root Circuit. In Proceedings of the 2021 34th International Conference on VLSI Design and 2021 20th International Conference on Embedded Systems (VLSID), Virtual Event, 20–24 February 2021; pp. 7–12.
39. Wu, D.; Miguel, J.S. In-Stream Stochastic Division and Square Root via Correlation. In Proceedings of the 2019 56th ACM/IEEE Design Automation Conference (DAC), Las Vegas NV, USA, 2–6 June 2019; pp. 1–6.
40. Qin, Z.; Qiu, Y.; Zheng, M.; Dong, H.; Lu, Z.; Wang, Z.; Pan, H. A universal approximation method and optimized hardware architectures for arithmetic functions based on stochastic computing. *IEEE Access* **2020**, *8*, 46229–46241. [CrossRef]
41. Gaines, B.R. R68-18 random pulse machines. *IEEE Trans. Comput.* **1968**, *100*, 410. [CrossRef]
42. Gaines, B.R. Stochastic computing systems. In *Advances in Information Systems Science*; Springer: Berlin/Heidelberg, Germany, 1969; pp. 37–172.
43. Kimball, J.W.; Krein, P.T.; Cahill, K.R. Modeling of capacitor impedance in switching converters. *IEEE Power Electron. Lett.* **2005**, *3*, 136–140. [CrossRef]
44. Forencich, A. Verilog Implementation of Mersenne Twister PRNG. 2018. Available online: https://github.com/alexforencich/verilog-mersenne (accessed on 15 February 2020).

 *technologies*

Parasitic Coupling in 3D Sequential Integration: The Example of a Two-Layer 3D Pixel [†]

Petros Sideris [1,2], Arnaud Peizerat [2], Perrine Batude [2], Gilles Sicard [2] and Christoforos Theodorou [1,*]

[1] University Grenoble Alpes, Univ. Savoie Mont Blanc, Centre National de la Recherche Scientifique (CNRS), Grenoble INstitut Polytechnique (Grenoble INP), Institut de Microélectronique Electromagnétisme Photonique-LAboratoire d'Hyperfréquences et de Caractérisation (IMEP—LAHC), 38000 Grenoble, France; psiderisd@gmail.com

[2] Commissariat à l'Energie Atomique et aux Energies Alternatives—Laboratoire d'Electronique des Technologies de l'Information (CEA—LETI), University Grenoble Alpes, 38054 Grenoble, France; arnaud.peizerat@cea.fr (A.P.); perrine.batude@cea.fr (P.B.); gilles.sicard@cea.fr (G.S.)

* Correspondence: christoforos.theodorou@grenoble-inp.fr

† This paper is an extended version of our paper published in 10th International Conference on Modern Circuits and Systems Technologies (MOCAST), Thessaloniki, Greece, 5–7 July 2021.

Abstract: In this paper, we present a thorough analysis of parasitic coupling effects between different electrodes for a 3D Sequential Integration circuit example comprising stacked devices. More specifically, this study is performed for a Back-Side Illuminated, 4T–APS, 3D Sequential Integration pixel with both its photodiode and Transfer Gate at the bottom tier and the other parts of the circuit on the top tier. The effects of voltage bias and 3D inter-tier contacts are studied by using TCAD simulations. Coupling-induced electrical parameter variations are compared against variations due to temperature change, revealing that these two effects can cause similar levels of readout error for the top-tier readout circuit. On the bright side, we also demonstrate that in the case of a rolling shutter pixel readout, the coupling effect becomes nearly negligible. Therefore, we estimate that the presence of an inter-tier ground plane, normally used for electrical isolation, is not strictly mandatory for Monolithic 3D pixels.

Keywords: image sensors; 3D pixels; 3D sequential integration; 3DSI; monolithic 3D; M3D; coupling; parasitic capacitances

Citation: Sideris, P.; Peizerat, A.; Batude, P.; Sicard, G.; Theodorou, C. Parasitic Coupling in 3D Sequential Integration: The Example of a Two-Layer 3D Pixel. *Technologies* **2022**, *10*, 38. https://doi.org/10.3390/technologies10020038

Academic Editors: Spiros Nikolaidis and Rodrigo Picos

Received: 11 January 2022
Accepted: 15 February 2022
Published: 28 February 2022

Publisher's Note: MDPI stays neutral with regard to jurisdictional claims in published maps and institutional affiliations.

1. Introduction

User-interactive applications are continuously emerging and driving the electronics industry towards the adoption of heterogeneous technologies in the sense that the analog sensing parts are integrated together with digital processing parts. The More-than-Moore technology development direction is a key enabler for such heterogeneous integrations, as it involves a wide variety of people–environment interaction applications [1].

An important driving application within the More-than-Moore scheme is the CMOS Image Sensor (CIS), because it is a circuit that requires the heterogeneous integration of different system parts: a photon-to-electron converter (photodiode) functions as the sensing interface in the pixel array and the readout part consists of an analog circuit that transmits information from the pixel to a digital circuit for processing. Moreover, CIS is an ideal candidate for studying coupling effects in 3D integration technologies, as it is highly sensitive to noise (dynamic) and mismatch (static) variations, while simultaneously being an extremely attractive application of 3D Integration for the semiconductor industry [2]. In particular, 3D CIS allows the development of smarter, more advanced sensors by co-integrating different blocks (Analog, Digital and RF) in multiple stacked tiers. By using the 3D stack integration process, the CMOS processing part can be stacked on top of the pixel, enabling the use of more advanced technology nodes for the processing circuit.

Although, until today, 3D stacking is predominantly used for 3D CIS, constraints concerning the alignment capabilities cannot allow more aggressive pixel miniaturization, which is required for future CIS generations [3,4]. However, in the case of 3D Sequential Integration (3DSI) [5,6], where one tier is processed on top of the other instead of being stacked, this drawback can be overcome, achieving pixel partitioning with state-of-the-art pixel pitch. In addition, 3DSI offers 3D contacts of outstanding high-density between tiers (up to 10^8 3D via/mm^2), enabling partitioning with high connectivity and low latency. Therefore 3DSI allows the co-integration of dense logic and memory layers but also heterogeneous components such as MEMS/NEMS for the compact coexistence of sensing and computing [7].

Despite the fact that the 3DSI technology approach offers great opportunities for the domain of CSI, it is also prone to many challenges, such as the limited Thermal Budget (TB) for the fabrication of the top-tier devices. To date, Low Temperature (LT) devices have been successfully fabricated and optimized for both low-voltage (LV) [5] and high-voltage (HV) [8] applications. Using such LT devices, Coudrain et al. [9] have investigated the feasibility of a 3DSI Back-Side Illuminated (BSI) CIS with miniaturized pixels, achieving a photodiode area increase by 44% for a 1.4 µm pitch. However, the other major challenge, concerning the impact of the inter-tier parasitic coupling on a 3DSI CIS performance, has not been examined in depth.

In this work, we present a thorough analysis of the possible coupling effects in the realization of a BSI 4–Transistor (4T) pixel with its diode and Transfer Gate (TG) on the bottom tier and the rest of its circuitry on the top tier of a 3DSI process, as an example of parasitic coupling analysis in a 3DSI circuit. In Section 2, the general principles and operation scheme of a CMOS imager are presented, including a presentation of the basic CIS architectures and the most critical performance metrics. In Section 3, the inter-tier coupling effects are analyzed, first at a single device level and then at the pixel level.

2. CMOS Imager: Architectures and Integration

2.1. CIS Standard Architectures and Operation

The most basic pixel architecture, called Passive Pixel Sensor (PPS) [10], contains passive pixels with no amplification, with only a photodiode for light detection and transistor switch for row selection, as shown in Figure 1a. Due to the lack of amplification or more sophisticated circuit, this architecture suffers from poor image quality, high KTC noise level and slow readout [11,12].

A major improvement to PPS architecture is the so-called Active Pixel Sensor (APS) [13], which incorporates an in-pixel amplifier for every pixel. Therefore, each pixel is composed of a photodiode (PD), a reset transistor (RST) and a source-follower (SF) amplifier, as demonstrated in Figure 1b. This architecture has reduced power consumption, random access and high-speed readout, thanks to the fact that the readout output is a voltage instead of charge transfer. On the downside, having additional transistors per pixel degrades the Fill-Factor. Finally, the issue of the kTC noise generated by the photodiode reset is not resolved.

In order to address the high reset noise issue, a Pinned Photodiode (PPD) pixel was introduced [14], resulting in the architecture shown in Figure 1c, which is the same as the APS one, but with the PPD connected to the readout circuit. This is achieved by an extra Transfer Gate (TX) and a Sense Node (SN). This 4T–APS architecture further allows the implementation of a fast Correlated Double Sampling (CDS) technique at the column level. Finally, thanks to the superior noise performance of the PPD [3], 4T–APS is currently the preferred architecture for CIS pixels in a variety of applications such as mobile imaging, digital still and video cameras, as well as surveillance cameras.

A schematic representation of the voltage output for a 3T–APS pixel is presented in Figure 2, showing how the operation sequence consists of three stages:

1.   Reset (RST ON): The photodiode voltage is set to a reference voltage V_{ref};

2. Exposure (RST OFF): The detected photons decrease the reverse photodiode voltage during the integration time (t_{int});
3. Readout (RS and SF ON): The output voltage level is sampled and further processed at the column level.

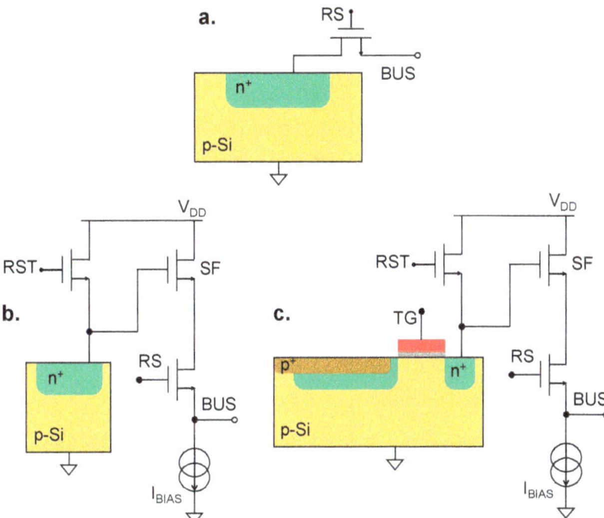

Figure 1. (**a**) Passive CMOS pixel based on a single in-pixel transistor, (**b**) active CMOS pixel based on an in-pixel amplifier and (**c**) active CMOS pixel based on an in-pixel amplifier in combination with a pinned photodiode.

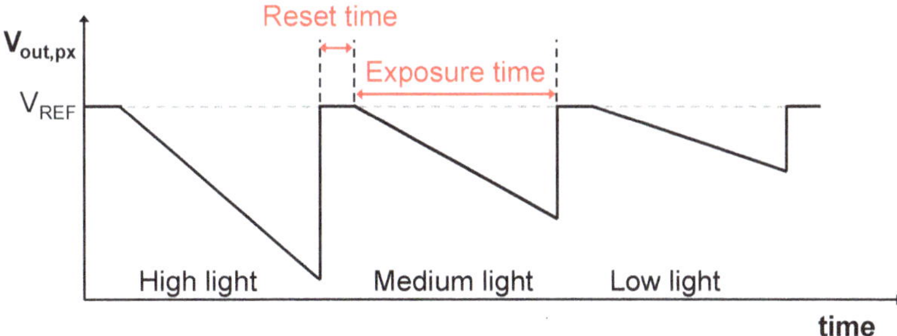

Figure 2. Operation principle of a 3T–APS pixel.

2.2. The Back-Side Illumination Integration Scheme

In the case of Front-Side Illumination, the optical path includes the total Back End Of Line (BEOL) thickness, inducing losses and crosstalk between pixels, because of the reflection on the metal lines. By flipping the sensor so that light drops directly on the photodiode without passing through the pixel's BEOL, the BSI integration scheme can be achieved. This is illustrated schematically in Figure 3, along with the example of stacked tiers of a 3D integration. The BSI scheme has become the preference for high-end consumer applications, dictated by the mobile phone market which requires continuously higher resolution for the same sensor size [3,4]. Currently, more than half of the mobile phone market utilize BSI integration [15]. Since 2010, the industrial trend is to combine BSI and 3D stacking/integration to reach ultimate performance, while at the same time maintaining a small pixel size.

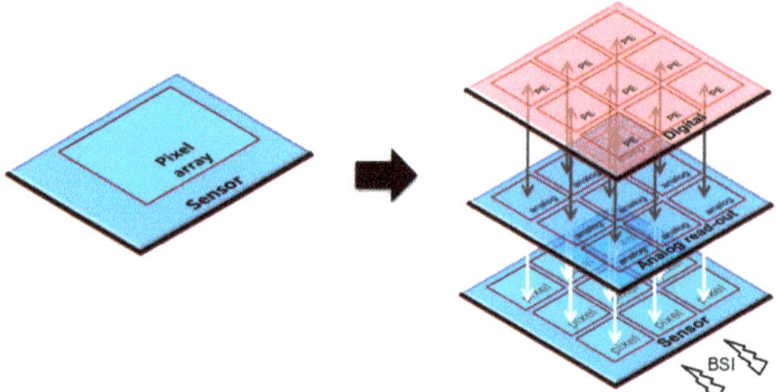

Figure 3. Realization of a CIS with planar (**left**) and three-dimensional partitioning (**right**).

2.3. CIS Performance Metrics

There are a series of important performance metrics for a CIS that are worth mentioning, before presenting our simulation study, such as the Signal-to-Noise Ratio (SNR), the Dynamic Range (DR) and the Conversion Gain (CG) amongst others.

SNR is defined as the ratio of the useful signal amplitude to the amplitude of undesirable noise and is a good indicator of image quality. Therefore it has to be maximized either by increasing the sensitivity and CG or by decreasing the noise floor [11,16]. SNR is reduced in the case of a smaller pixel area, which means less absorbed photons for the same duration. However, the reduction in pixel size provides higher spatial resolution, at a lower light sensitivity as a trade-off. Minimizing the pixel size will also affect the Full Well Capacity (FWC) and, thus, the DR (these parameters will be below).

The Fill Factor (FF) is another parameter that is affected by pixel scaling [11,16]. It is defined as the percentage of the photosensitive area with regards to the total pixel area. Thus, it expresses what portion of the total pixel area is utilized for photon collection. A high FF value means higher sensitivity; thus, it ought to be maximized, for example, by downscaling pixel transistors or by adding microlenses to guide the light towards the photodiode area.

Regarding the sensitivity of a sensor with linear response, as is the case of 3T/4T–APS, it is equal to the slope of the transfer function (in V/lux.s or e^-/lux.s). It corresponds to the change in output potential for a given light intensity and integration time. It is highly dependent on the Quantum efficiency (QE) of the sensor [11], a quantity that shows how efficiently the photons are collected and converted to electrons. One method to maximize it is to use anti-reflecting coating or optimize the stack between the sensor surface and the photodiode, or to use BSI, in order to avoid reflections at the interfaces.

The DR of a CIS is defined as the range of light intensity that can be measured with no distortion by the sensor [11,16]. It can be calculated as $DR = 20\log(S_{max}/S_{min})$, where S_{max} is the highest detectable signal and S_{min} the lowest one (essentially the noise floor). On the other hand, S_{max} is limited by the Full Well Capacity and pixel saturation.

An important parameter for our study is CG, which is defined as $CG = \Delta V_{out}/N_e$, where ΔV_{out} is the pixel output ($V_{out,int} - V_{out,ref}$) when the number of electrons is equal to N_e in a single packet [11,16]. N_e depends on photon flux and QE. It is measured in V/e^- and it characterizes the charge-to-voltage conversion; thus, a high CG results in higher sensitivity, especially at low light. Another method to express CG is through (X), taking into account SN capacitance, C_{SN}, and any additional parasitic capacitance, C_P.

$$CG = \frac{q}{C_{SN} + C_P} \tag{1}$$

Finally, the amount of charge that can be detected without reaching saturation is expressed through the FWC metric, which is measured in number of charges and determines the sensor's DR [11,16]. If we neglect the noise, FWC can be roughly approximated by FWC = $qC_{PD}V_{PD}$, where V_{PD} is the applied voltage across the photodiode, and C_{PD} is the photodiode capacitance. By increasing C_{PD}, therefore, one can directly increase the sensor's FWC. However, in that case, CG would be decreased due to the increased capacitance. This in turn may result in a range decrease at the low intensity end, and this contradiction is actually the well-known DR/sensitivity tradeoff.

3. Parasitic Capacitance Coupling in a Two-Layer 3DSI Pixel

As already mentioned in the Introduction, pixel partitioning in two layers of a 3DSI process is a very promising technique to boost an imager's performance by increasing the photodiode area ratio, and it has dedicated layers for each type of circuit (read-out, digital etc.). Nevertheless, positioning transistors right above the photodiode's transfer gate at a submicrometre distance introduces a high risk of parasitic capacitance coupling, which will be investigated in this Section. At first, in order to be certain if there can be a significant coupling-induced threshold voltage shift and, if so, to quantify it, we performed simulations at a device level, without taking into account the particularities of a pixel's circuit operation and chronogram. Afterwards, in Section 3.2, we present the simulation results we obtained at a circuit level, examining in which cases the parasitic coupling can affect or not the pixel's output precision.

3.1. Investigation at Device Level
3.1.1. Simulated Structure Details

In order to carry out our study, the simulation structure depicted in Figure 4 was considered. As shown in the cross-section of Figure 4a, our setup has its pixels sequentially integrated in such a manner that PPD and TG are placed at the bottom layer and the rest of the readout circuitry is placed right above them, with an Inter-Layer Dielectric (ILD) of 200 nm thickness separating them. TG can toggle between 0 and V_{DD}, enabling photo-generated electrons to be transferred from the photodiode area to SN and then to the drain of RST and the gate of the SF via a 3D contact.

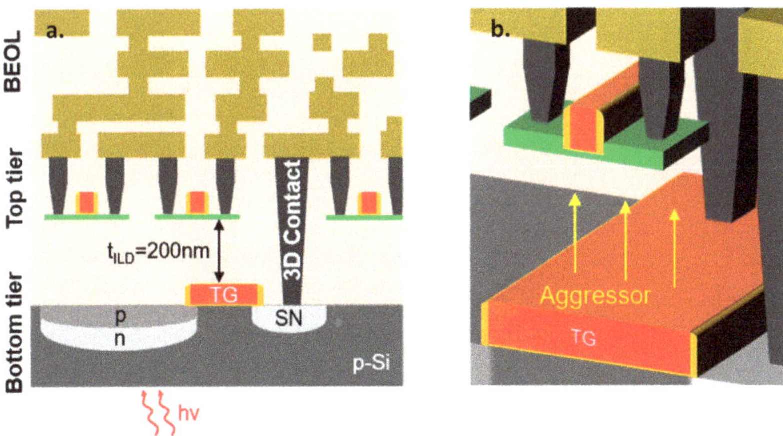

Figure 4. (**a**) Cross-section of 4T pixel, partitioned in 3DSI. (**b**) In the most critical case, the top device is placed right above the TG electrode carrying a voltage that can go up to V_{DD} = 2.5 V.

In order to extend the voltage swing of top-tier devices, analog devices with power supply of 2.5 V were considered for the top layer. In addition, the most critical condition of inter-tier coupling has been chosen by placing the TG electrode right under each one of the top layer devices, as shown in Figure 4b.

3.1.2. Impact of Inter-Tier Coupling on Electrical Parameters

In order to assess whether inter-tier static coupling can be detrimental to the functionality of the 3DSI pixel, we investigated the impact of the capacitive coupling of the TG placed at the bottom tier on each top device performance. By varying the TG gate voltage bias within its normal operation limits (0–2.5 V), we observe a shift of the I_D–V_G characteristics for the top devices, as shown in Figure 5a. This behavior can be attributed to the fact that the top devices are actually asymmetrical SOI structures, where ILD plays the role of the Buried Oxide (BOX). It is well-known in SOI devices [17] that this effect is nonlinear, with its maximum in weak inversion and equal to a constant value when the device enters strong inversion.

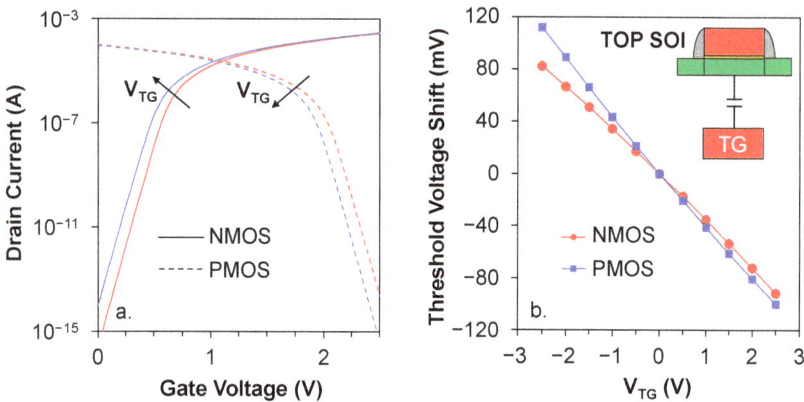

Figure 5. (**a**) Impact of TG coupling (V_{TG} = 0 (red) −2.5 V (blue)) on the input characteristics for NMOS and PMOS devices. (**b**) Extracted threshold voltage shift versus TG voltage bias (−2.5 V to 2.5 V with 0.5 V step).

Figure 5b shows the extracted threshold voltage shift (ΔV_{TH}) versus V_{TG} (from −2.5 V to +2.5 V) and a nearly linear relation was obtained, which allowed the extraction of the back-bias efficiency γ for the top devices (35 mV/V for NMOS and 42 mV/V for PMOS). V_{TH} was extracted using the constant current method [18] for each I_D–V_G curve.

In order to evaluate the strength of TG-induced static coupling on the top device performance, we benchmarked it against a temperature variation of 100 °C, which is a typical range in consumer electronics. Contrary to capacitive coupling, the effect of temperature varies depending on the gate voltage V_G. Indeed, our simulations (Figure 6) show that when proceeding from 253 K to 353 K (−20 °C to 80 °C), an increase in the leakage current and a decrease in ON current were observed, whereas at a specific gate voltage around V_{TH}, the drain current remains unaffected by T. This effect is well known in the literature and is due to the canceling out between the rise in carrier concentration with temperature for low V_G and the decrease in carrier mobility at high V_G [19]. The voltage at which this happens is characterized as the Zero Temperature Coefficient (ZTC) point [20].

Eventually, the I_D–V_G shift resulting from both coupling and temperature variation can be translated in an alteration of the top device electrical parameters, namely the V_{TH}, the leakage current (I_{OFF}) and the saturation current (I_{ON}). The comparison of the extracted parameters for the two effects is shown in Figures 7 and 8. The results show that they are approximately in the same order of magnitude while the limited ΔV_{TH} due to the temperature variation is attributed to the ZTC point near V_{TH}. Moreover, as observed in Figure 7, I_{OFF} is significantly shifted with V_{TG}, which can be detrimental for memory blocks comprising switch transistors, placed at the top-tier above TG.

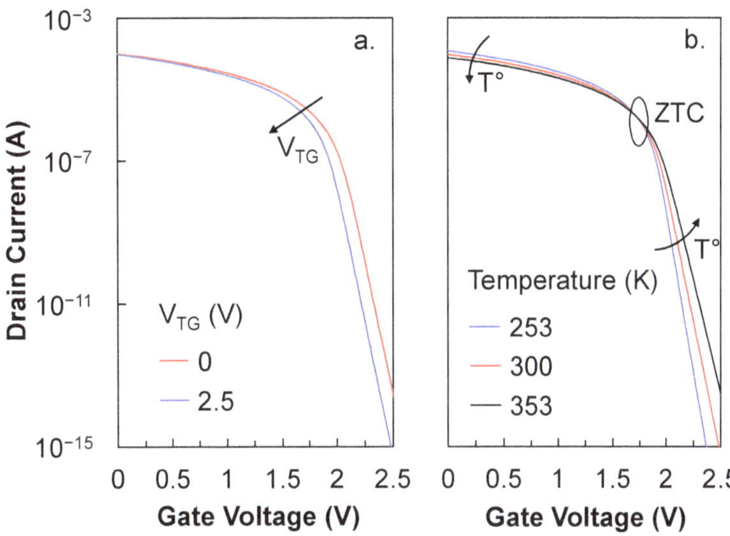

Figure 6. Impact of coupling (**a**) and temperature variation (**b**) on the top-tier SOI input characteristics.

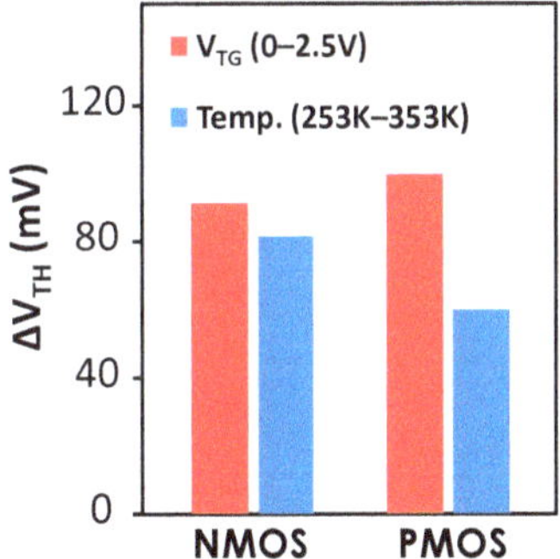

Figure 7. Comparison of the V_{TH} variation of the top-tier NMOS and PMOS due to TG coupling or due to a 100 K temperature increment.

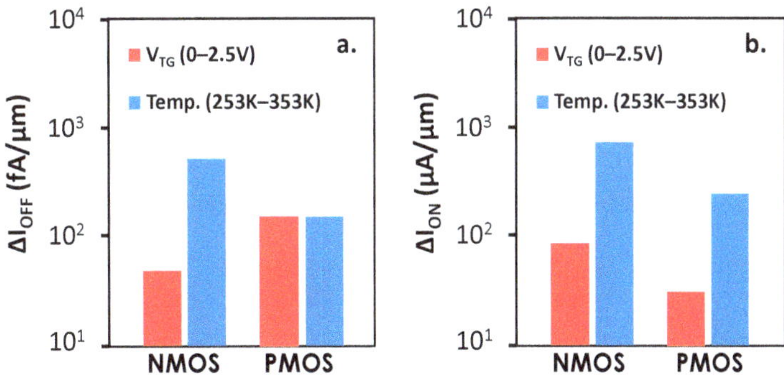

Figure 8. Comparison of the I_{OFF} (**a**) and I_{ON} (**b**) variation of the top-tier NMOS and PMOS due to TG coupling or due to a 100 K temperature increment.

3.2. Investigation at Pixel Level

3.2.1. Pixel Topology and Chronogram

In order to carry out our analysis at the pixel level, the 4T–APS topology was selected, which consisted of an NMOS–TG and the RST, SF and RS in a PMOS circuit configuration, as illustrated in Figure 9a. As shown from the chronogram of the 4T–APS pixel readout operation presented in Figure 9b, during the readout cycle, TG was switched ON following the SN reset, allowing the diffusion of photo-generated electrons, which in turn cause a voltage drop at the input of the SF. Due to the rough unity gain of SF, the voltage drop is transferred nearly at the same level at its output.

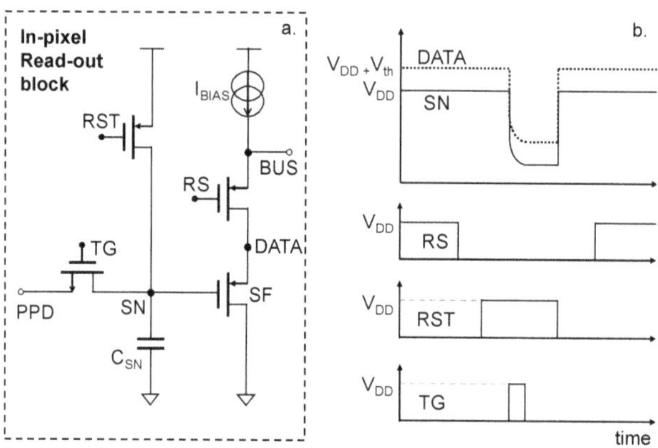

Figure 9. (**a**) 4T pixel readout circuit. (**b**) chronogram of a readout cycle.

The efficiency of the aforementioned operation is characterized by the CG in $\mu V/e^-$, given by the following [21]:

$$CG = \frac{q G_{SF}}{C_{SN} + C_{GD} + (1 - G_{SF})C_{GS}},\tag{2}$$

where q is the elementary charge, G_{SF} is the SF gain, C_{SN} is the sum of parasitic capacitances at the SN node and C_{GS} and C_{GD} are the gate to source and gate to drain capacitances of SF.

The gain G_{SF} of the SF, on the other hand, is expressed as follows:

$$G_{SF} = \frac{g_{m,SF}}{g_{ms,SF}} = \frac{1}{n},\tag{3}$$

where $g_{m,SF}$ and $g_{ms,SF}$ are the gate and source transconductances, respectively, and n is the body factor of SF. The gain is approximately equal to unity in the case where the BG of SF can be tied to the source or else it is process-dependent and is given by $n = 1 + \gamma$, where γ is the back-bias efficiency. With the increase in ILD thickness, n approaches unity. For the γ values extracted in the previous section, we have evaluated the gain G_{SF} of the NMOS and PMOS devices as 0.97 and 0.96, respectively.

The cutoff frequency (f_c) of the SF is given by the following [21]:

$$f_c = \frac{g_{m,SF}}{2\pi \cdot C_{out,SF} \cdot G_{SF} \cdot (C_{SN} + C_{GD} + C_{GS})},\tag{4}$$

where $C_{out,SF}$ is the capacitance observed at the source of the SF that is the column-level capacitance if there are no other stages in between.

3.2.2. Impact of TG Coupling on Pixel Electrical Parameters

In order to evaluate 3DSI impacts on the two critical parameters CG and f_c analyzed above, parasitic extraction was performed concerning a single- and a two-layer implementation of our pixel, as illustrated in Figure 10, and the results along with CG values are presented in Table 1. As observed, the sums of C_{SN} and C_{GD}, as well as C_{GS}, slightly increased by 48aF and 44aF, respectively, which may be attributed to the proximity of the top tier.

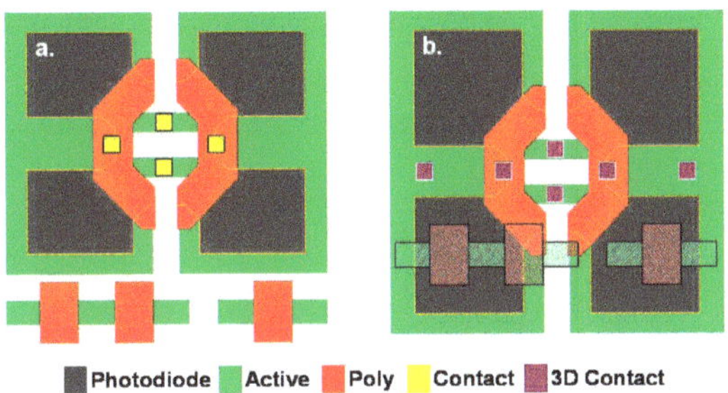

Figure 10. Shared pixels at photodiode layer: 2D (**a**) vs. 3D (**b**). Photodiode area (in grey) increased by 44% when the three readout transistors were placed at the top tier.

Table 1. Two-dimensional vs. three-dimensional parasitic capacitances and CG.

	C_{SN} + C_{GD} (fF)	C_{GS} (fF)	$C_{out,SF}$ (pF)	CG (μV/e$^-$)
2D	4.432	1.093	2	34.319
3D	4.48	1.137	2	33.942

Furthermore, the column-level capacitance reveals an even smaller increase in 10aF, resulting in a minor difference in the conversion gain (ΔCG = 0.377 μV/e$^-$) and the AC response of the two-layer pixel (Δf_c = 0.244 Hz) compared to the single-layer one. The low 3DSI impact on the CG also suggests that noise performance will not be degraded.

The diagram of Figure 11 shows the output voltage of SF versus the number of photogenerated electrons at SN, where it is apparent that there is a constant vertical shift of the response without significant change in its slope (CG remains the same) while varying TG bias. The former can be considered as an offset that can be easily adjusted during the readout process.

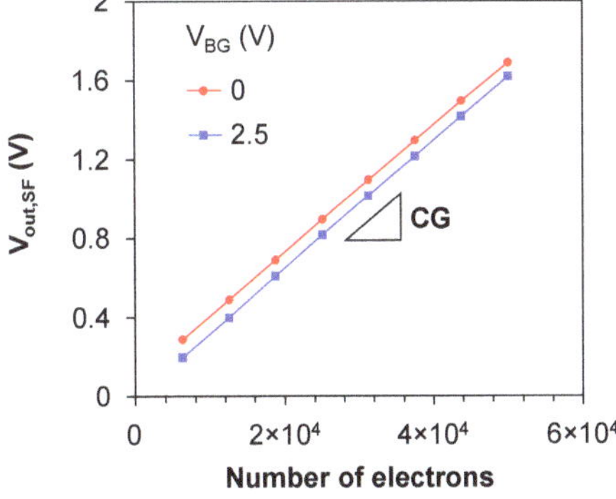

Figure 11. Output voltage of the in-pixel SF transistor versus the number of photo generated electrons at SN for the voltage bias limits of the bottom tier TG (0 V–2.5 V). The slope provides that CG was not altered with TG coupling.

Concerning the rest of pixel performance metrics, they are either related to the photodiode technology and size (similarly to FWC and FF) and, thus, are not affected by 3DSI layering, or directly related to the CG, such as the SNR and DR, which was found almost unchanged by parasitic coupling.

Continuing our analysis, in order to estimate the impact of the inter-tier coupling on the transient response of the top readout circuit, the two scenarios of Figure 12 were employed. The first is considered when SF is placed above a TG of the same pixel (Figure 12a). In that case, charge transfer occurs right at the time in which TG is enabled, as shown in the pixel readout cycle. Thus, sampling processes performed right before and after the charge transfer cannot result in an error.

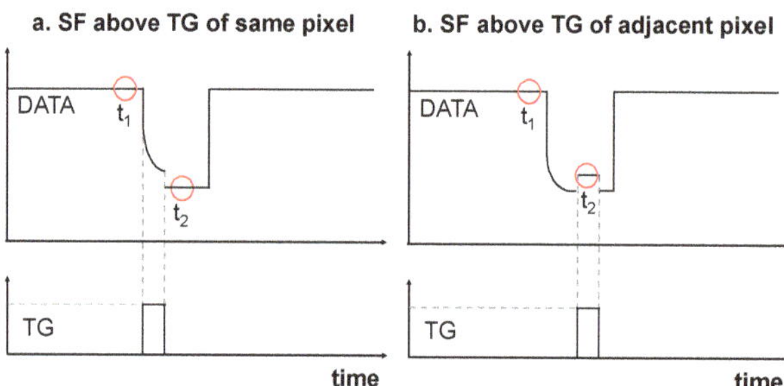

Figure 12. (**a**) For SF above a TG of the same pixel, TG switches ON during the transfer of e^- from PPD to the SN. Sampling is, thus, performed at t_1 and t_2 without readout errors. (**b**) In the scenario of an SF placed above a TG of an adjacent pixel, sampling can contain erroneous value due to TG coupling.

On the contrary, our second scenario considers an SF placed above the TG of an adjacent pixel (Figure 12b). In this case, TG is not synchronized with the pixel readout; hence, sampling can contain erroneous information. This sampling error $\Delta V_{out,SF}$ has been extracted versus the number of the photogenerated electrons at the SN and for various bias currents of the SF, with the results shown in Figure 13a. In order to evaluate the strength of this error, a comparison was made with the readout error resulting in temperature variations (253 K–353 K), which is presented in Figure 13b. It is evident that the coupling-induced readout error is significant compared to the temperature-induced one, especially for low light conditions and low bias current.

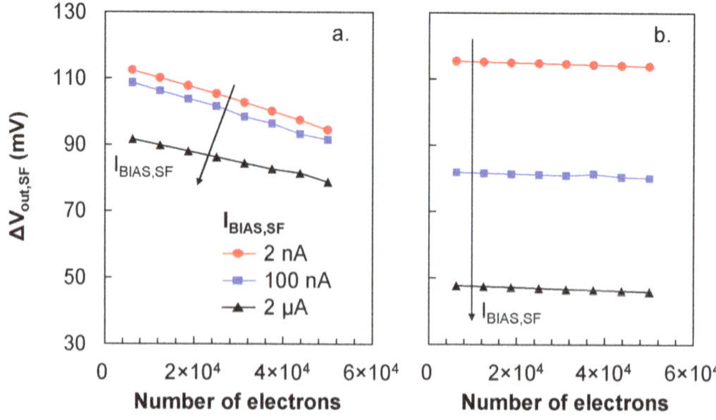

Figure 13. Readout error of the in-pixel SF transistor versus the number of photo generated electrons at SN resulting from (**a**) SF–TG coupling of Figure 7b and (**b**) sensor temperature variations.

3.2.3. Inter-Tier Ground Plane Necessity

Recent studies [22,23] have demonstrated efficient decoupling solutions between sensitive tiers in 3DSI via the process integration of a conductive layer, i.e., an inter-tier Ground Plane (GP).

Depending on each application's sensitivity and functionality, an inter-tier Ground Plane integration must be considered, taking severely into account process complexity. Nevertheless, in cases where an effective isolation between tiers in 3DSI is required, studies show that the integration of an inter-tier GP made of polysilicon can offer a reduction in vertical static coupling by five orders of magnitude [23].

Concerning the example of the two-layer 3DSI pixel studied in this work, in a typical rolling readout operation, sequential pixel activation implies that there is no probability for a readout error due to TG coupling. Consequently, TG will be enabled outside the readout cycle of this pixel and, thus, does not interfere. Furthermore, the Correlated Double Sampling (CDS) stage that exists commonly after the readout circuit eliminates possible readout errors. Therefore, the direct stacking of the readout tier upon the photodiode area is safe without the necessity for electrical isolation in a 3DSI CIS. However, an inter-tier GP is mandatory in cases where sensitive blocks are considered to be placed above the photodiode area, such as in-pixel frame memory for which the leakage current is a critical parameter.

4. Conclusions

To summarize, we have presented an investigation of coupling-induced effects in a 3DSI PMOS pixel with the aid of TCAD simulations. Our results show that coupling from TG can cause an electrical parameter variation as important as the one induced by a 100 degrees temperature variation. For switch transistors, where the leakage is a critical parameter, this could be a very limiting effect. However, for the in-pixel SF transistor, we demonstrated that the impact of inter-tier electrical coupling on the CG and the AC performance is negligible. Concerning the rest of the indicators presented in Section 2.3, none of them should be affected (especially QE and FWC since they only depend on the photodiode technology). Hence, in addition to the very slight increase in CG, pixel performance should be maintained.

We have further shown that SF–TG coupling may cause a readout error if SF is placed above the TG of a nearby pixel, i.e., in the case that it is not synchronized with the pixel readout. This is not necessarily a limitation for the CIS performance, because if the readout is performed following the rolling shutter scheme, pixel activation is sequential. Therefore, despite strong electrical coupling and high threshold voltage shifts (~100 mV) for top-tier devices, we demonstrated that a sequentially integrated 3D CIS can have an inherent immunity to inter-tier coupling, with zero readout errors.

Author Contributions: Conceptualization, P.S., P.B., G.S. and C.T.; methodology, P.S., A.P., G.S. and C.T.; software, P.S.; validation, P.S. and A.P.; formal analysis, P.S., A.P. and G.S.; investigation, P.S., A.P. and G.S.; resources, P.B.; data curation, P.S.; writing—original draft preparation, P.S. and C.T.; writing—review and editing, A.P., P.B. and G.S.; visualization, P.S.; supervision, A.P., G.S. and C.T.; project administration, P.B.; funding acquisition, P.B., G.S. and C.T. All authors have read and agreed to the published version of the manuscript.

Funding: This work was partially supported by the H2020 3DMUSE European project (Funding number: 780548).

Institutional Review Board Statement: Not applicable.

Informed Consent Statement: Not applicable.

Data Availability Statement: Data will be available upon publication.

Acknowledgments: The authors would like to thank Josep Segura for fruitful discussions.

Conflicts of Interest: The authors declare no conflict of interest.

References

1. Arden, W.; Brillouët, M.; Cogez, P.; Graef, M.; Huizing, B.; Mahnkopf, R. "More-than-Moore" White Paper; 2010; Version 2. Available online: http://www.itrs2.net/uploads/4/9/7/7/49775221/irc-itrs-mtm-v2_3.pdf (accessed on 2 January 2022).
2. Haruta, T.; Nakajima, T.; Hashizume, J.; Umebayashi, T.; Takahashi, H.; Taniguchi, K.; Kuroda, M.; Sumihiro, H.; Enoki, K.; Yamasaki, T.; et al. A 1/2.3inch 20Mpixel 3-layer stacked CMOS Image Sensor with DRAM. In Proceedings of the Digest of Technical Papers—IEEE International Solid-State Circuits Conference, San Francisco, CA, USA, 5–9 February 2017; Institute of Electrical and Electronics Engineers Inc.: Piscataway, NJ, USA, 2017; Volume 60, pp. 76–77.
3. Theuwissen, A. CMOS image sensors: State-of-the-art and future perspectives. In Proceedings of the ESSDERC 2007-37th European Solid State Device Research Conference, Munich, Germany, 11–13 September 2007; pp. 21–27. [CrossRef]
4. Coudrain, P.; Magnan, P.; Batude, P.; Gagnard, X.; Leyris, C.; Vinet, M.; Castex, A.; Lagahe-Blanchard, C.; Pouydebasque, A.; Cazaux, Y.; et al. Investigation of a sequential three-dimensional process for back-illuminated CMOS image sensors with miniaturized pixels. *IEEE Trans. Electron Devices* **2009**, *56*, 2403–2413. [CrossRef]
5. Brunet, L.; Batude, P.; Fenouillet-Beranger, C.; Besombes, P.; Hortemel, L.; Ponthenier, F.; Previtali, B.; Tabone, C.; Royer, A.; Agraffeil, C.; et al. First demonstration of a CMOS over CMOS 3D VLSI CoolCube^{TM} integration on 300mm wafers. In Proceedings of the 2016 IEEE Symposium on VLSI Technology, Honolulu, HI, USA, 14–16 June 2016; IEEE: Piscataway, NJ, USA, 2016; pp. 1–2.
6. Batude, P.; Brunet, L.; Fenouillet-Beranger, C.; Andrieu, F.; Colinge, J.-P.; Lattard, D.; Vianello, E.; Thuries, S.; Billoint, O.; Vivet, P.; et al. 3D Sequential Integration: Application-driven technological achievements and guidelines. In Proceedings of the 2017 IEEE International Electron Devices Meeting (IEDM), San Francisco, CA, USA, 2–6 December 2017; IEEE: Piscataway, NJ, USA, 2017; pp. 3.1.1–3.1.4.
7. Vivet, P.; Sicard, G.; Millet, L.; Chevobbe, S.; Chehida, K.B.; Angel Cubero, L.; Alegre, M.; Bouvier, M.; Valentian, A.; Lepecq, M.; et al. Advanced 3D Technologies and Architectures for 3D Smart Image Sensors. In Proceedings of the 2019 Design, Automation & Test in Europe Conference & Exhibition (DATE), Florence, Italy, 25–29 March 2019; IEEE: Piscataway, NJ, USA, 2019; pp. 674–679.
8. Cavalcante, C.; Garros, X.; Batude, P.; Tataridou, A.; Lacord, J.; Casse, M.; Theodorou, C.; Karatsori, T.; Gassilloud, R.; Fenouillet-Beranger, C.; et al. Low temperature high voltage analog devices in a 3D sequential integration. In Proceedings of the 2020 International Symposium on VLSI Technology, Systems and Applications, VLSI-TSA 2020, Hsinchu, Taiwan, 10–13 August 2020.
9. Coudrain, P. Contribution au développement d'une technologie d'intégration tridimensionnelle pour les capteurs d'images CMOS à pixels actifs. Ph.D. Thesis, Institut Supérieur de l'Aéronautique et de l'Espace (ISAE), Toulouse, France, 2009.
10. Weckler, G.P. Operation of p-n Junction Photodetectors in a Photon Flux Integrating Mode. *IEEE J. Solid-State Circuits* **1967**, *2*, 65–73. [CrossRef]
11. El Gamal, A.; Eltoukhy, H. CMOS image sensors. *IEEE Circuits Devices Mag.* **2005**, *21*, 6–20. [CrossRef]
12. Zhao, C.; Kanicki, J.; Konstantinidis, A.C.; Patel, T. Large area CMOS active pixel sensor x-ray imager for digital breast tomosynthesis: Analysis, modeling, and characterization. *Med. Phys.* **2015**, *42*, 6294–6308. [CrossRef] [PubMed]
13. Yadid-Pecht, O.; Ginosar, R.; Diamand, Y.S. A Random Access Photodiode Array for Intelligent Image Capture. *IEEE Trans. Electron Devices* **1991**, *38*, 1772–1780. [CrossRef]
14. Guidash, R.M.; Lee, T.H.; Lee, P.P.K.; Sackett, D.H.; Drowley, C.I.; Swenson, M.S.; Arbaugh, L.; Hollstein, R.; Shapiro, F.; Domer, S. 0.6 μm CMOS pinned photodiode color imager technology. In Proceedings of the Technical Digest—International Electron Devices Meeting, IEDM, Washington, DC, USA, 10 December 1997; pp. 927–929.
15. Yole Developpement Status of the CMOS Image Sensor Industry 2021. Available online: https://s3.i-micronews.com/uploads/2021/08/YINTR21167-Status-of-the-CMOS-Image-Sensor-Industry-2021-Sample.pdf (accessed on 2 January 2022).
16. Kadura, L. *New FDSOI-Based Integrated Circuit Architectures Sensitive to Light for Imaging Applications*; Grenoble Alpes University: Grenoble, France, 2019.
17. Amara, A.; Rozeau, O. *Planar Double-Gate Transistor*; Amara, A., Rozeau, O., Eds.; Springer: Dordrecht, The Netherlands, 2009; ISBN 978-1-4020-9327-2.
18. Tsuno, M.; Suga, M.; Tanaka, M.; Shibahara, K.; Miura-Mattausch, M.; Hirose, M. Physically-based threshold voltage determination for MOSFET's of all gate lengths. *IEEE Trans. Electron Devices* **1999**, *46*, 1429–1434. [CrossRef]
19. Groeseneken, G.; Colinge, J.P.; Maes, H.E.; Alderman, J.C.; Holt, S. Temperature Dependence of Threshold Voltage in Thin-Film SOI MOSFET's. *IEEE Electron Device Lett.* **1990**, *11*, 329–331. [CrossRef]
20. Khanna, V.K. Temperature dependence of electrical characteristics of silicon MOS devices and circuits. In *Extreme-Temperature and Harsh-Environment Electronics Physics, Technology and Applications*; IOP Publishing: Bristol, UK, 2017.
21. Fowler, B. *Single Photon CMOS Imaging Through Noise Minimization*; Seitz, P., Theuwissen, A.J., Eds.; Springer Series in Optical Sciences; Springer: Berlin/Heidelberg, Germany, 2011; Volume 160, pp. 159–195; ISBN 978-3-642-18442-0.
22. Vandooren, A.; Wu, Z.; Khaled, A.; Franco, J.; Parvais, B.; Li, W.; Witters, L.; Walke, A.; Peng, L.; Rassoul, N.; et al. Buried metal line compatible with 3D sequential integration for top tier planar devices dynamic Vth tuning and RF shielding applications. *Dig. Technol. Pap.-Symp. VLSI Technol.* **2019**, *2019*, T56–T57. [CrossRef]
23. Sideris, P.; Andrieu, F.; Colinge, J.P.; Ghibaudo, G.; Theodorou, C.; Lugo-Alvarez, J.; Batude, P.; Brunet, L.; Acosta-Alba, P.; Kerdiles, S.; et al. Inter-tier Dynamic Coupling and RF Crosstalk in 3D Sequential Integration. *Technol. Dig.-Int. Electron Devices Meet. IEDM* **2019**, *2019*, 3.4.1–3.4.4. [CrossRef]

technologies

MDPI

Article

Lightweight Neural Network for COVID-19 Detection from Chest X-ray Images Implemented on an Embedded System [†]

Theodora Sanida *, Argyrios Sideris, Dimitris Tsiktsiris and Minas Dasygenis

Department of Electrical and Computer Engineering, University of Western Macedonia, 50131 Kozani, Greece; asideris@uowm.gr (A.S.); dtsiktsiris@uowm.gr (D.T.); mdasyg@ieee.org (M.D.)
* Correspondence: thsanida@uowm.gr; Tel.: +30-24610-56534
† This paper is an extended version of our paper published in 10th International Conference on Modern Circuits and Systems Technologies (MOCAST), Thessaloniki, Greece, 5–7 July 2021.

Abstract: At the end of 2019, a severe public health threat named coronavirus disease (COVID-19) spread rapidly worldwide. After two years, this coronavirus still spreads at a fast rate. Due to its rapid spread, the immediate and rapid diagnosis of COVID-19 is of utmost importance. In the global fight against this virus, chest X-rays are essential in evaluating infected patients. Thus, various technologies that enable rapid detection of COVID-19 can offer high detection accuracy to health professionals to make the right decisions. The latest emerging deep-learning (DL) technology enhances the power of medical imaging tools by providing high-performance classifiers in X-ray detection, and thus various researchers are trying to use it with limited success. Here, we propose a robust, lightweight network where excellent classification results can diagnose COVID-19 by evaluating chest X-rays. The experimental results showed that the modified architecture of the model we propose achieved very high classification performance in terms of accuracy, precision, recall, and f1-score for four classes (COVID-19, normal, viral pneumonia and lung opacity) of 21.165 chest X-ray images, and at the same time meeting real-time constraints, in a low-power embedded system. Finally, our work is the first to propose such an optimized model for a low-power embedded system with increased detection accuracy.

Keywords: classification; COVID-19; lightweight neural network; deep learning; medical imaging; chest X-rays

Citation: Sanida, T.; Sideris, A.; Tsiktsiris, D.; Dasygenis, M. Lightweight Neural Network for COVID-19 Detection from Chest X-ray Images Implemented on an Embedded System. *Technologies* **2022**, *10*, 37. https://doi.org/10.3390/technologies10020037

Academic Editors: Spiros Nikolaidis and Rodrigo Picos

Received: 15 December 2021
Accepted: 21 February 2022
Published: 25 February 2022

Publisher's Note: MDPI stays neutral with regard to jurisdictional claims in published maps and institutional affiliations.

1. Introduction

COVID-19 is a highly contagious and infectious disease that causes severe infection of the lower respiratory system. COVID-19 was officially declared a global pandemic by the World Health Organization (WHO) on 11 March 2020 [1]. The total number of COVID-19 cases worldwide up to this point, which changes almost every minute, is over 328 million, including 5 million deaths and 55 million active patients [2]. The increasing number of deaths from COVID-19 and the rapid rate of disease transmission makes it one of the world's most significant and most serious public health issues.

The clinical findings of COVID-19 infection are those of bronchopneumonia, where cough, dyspnea, fever and respiratory failure with acute respiratory distress syndrome (ARDS) occur [3,4]. Radiographic imaging is the fastest, cheapest, and most essential diagnostic tool for detecting pneumonia while providing lower patient radiation than computed tomography (CT) and magnetic resonance imaging (MRI) [5,6].

In COVID-19 disease, the lungs are affected resulting in severe viral pneumonia [7–9]. Deaths from viral pneumonia due to COVID-19 are constantly increasing day by day. Thus, a valid diagnosis of COVID-19 pneumonia is vital but often a difficult task due to the similar symptoms and radiological findings that occur between patients infected with COVID-19 and patients with viral pneumonia [10–12]. This difficulty can be alleviated by enhancing the arsenal of health workers using artificial intelligence in the form of machine learning.

In recent years, artificial intelligence (AI) has been established as a powerful tool in the field of medical services for classification, prediction, diagnosis, detection and segmentation [13–18]. In medical imaging, DL in chest X-rays has been shown to play an essential role in the detection of viral pneumonia, bacterial pneumonia and other chest diseases [19,20]. With the rapid spread of the virus and the fight against pandemics, researchers are using state-of-the-art DL techniques to classify COVID-19 X-rays [21–24]. Therefore, the design and development of DL models for the classification of COVID-19 infected radiographs is an urgent need to address the pandemic to make appropriate clinical decisions and significantly reduce workload [25].

The main goal of this study is to propose a robust, lightweight neural network, enhancing the work in [26], that provides high accuracy of chest X-ray prediction for COVID-19 detection executed on a low-power embedded system. The classic deep CNNs such as VGG16, VGG19 [27] have excellent performance, but they are difficult to deploy on devices with low hardware, due to the amount of model parameters and computational complexity. We use low-power embedded systems to design a low-cost portable point-of-care medical device that would be affordable for any hospital to acquire [28,29].

Many models have been developed and modified, but there is still room for improvement, as our research proves.

The main contributions of the proposed study can be summarized as follows:

- We propose a modified neural network structure of the MobileNetV2 model, to maximize the learning ability for classification of chest X-rays. The modified version of our architecture requires significant less training time than other existing DL architectures due to the small number of network parameters.
- Our design can classify four different categories of chest X-rays (COVID-19, normal, viral pneumonia and lung opacity). The accuracy of our approach is significantly higher than standard architecture and surpasses other state-of-the-art methods.
- This is the first study to investigate a large set of chest X-ray images (21.165 chest X-ray images) combined from many other studies that appear in the literature, which include few X-ray samples and mainly concern binary classifications.
- The modified structure of the architecture yields excellent classification results and, in combination with the small size of the network, leads to an attractive model for diagnosing chest X-rays in embedded systems.

The rest of the paper is organized as follows. In Section 2, we summarize related work. Subsequently, in Section 3 we give an outline of the procedure followed for the proposed approach, while in Section 4 we present the experimental results, and in Section 5 we conclude the article.

2. Related Work

In the global fight against the virus, the rapid and valid assessment of COVID-19 infected patients is a major challenge for the research community. Several studies have been performed to demonstrate the potential of DL in detecting COVID-19 infected chest X-rays in the field of medical imaging.

Kamal et al. [30] evaluated eight pre-trained CNN models (MobileNetV2, MobileNet, VGG19, ResNet50, ResNet50V2, DenseNet121, InceptionV3 and NasNetMobile). The aperture in the preformed plates of 744 images exposes 186 normal, 186 COVID-19, 186 bacterial pneumonia, and 186 viral pneumonia. Each DL model trained 1000 epochs. The results showed that for 4 class classification, the MobileNetV2 model achieved 95.40% accuracy. In [31] the authors proposed a model called Bayesian CNN. The used dataset consisted of 68 COVID-19, 1.583 normal, 2.786 bacterial pneumonia and 1.504 viral pneumonia images. Their CNN model was trained for 25 epochs and for 4 class classification achieved 89.82% accuracy. Khan et al. [32] proposed a model called CoroNet, which is based on the Xception model and consists of 33 million parameters. The used dataset in their tests consisted of 284 COVID-19, 310 normal, 330 bacterial pneumonia and 327 viral pneumonia chest radiography images. Their DL model was trained for 80 epochs and for 4 class classification (COVID-19, bacterial pneumonia, viral pneumonia, normal) achieved an accuracy of 89.60%.

Mahmud et al. [33] suggest a model called CovXNet. The used dataset consisted of 305 COVID-19, 305 normal 305 viral pneumonia and 305 bacterial pneumonia. Their CNN model was trained for 150 epochs and for 4 class classification achieved 90.20% accuracy. Wang et al. [34] proposed a model called COVID Net, which consists of 11.75 million parameters. The dataset they used in their tests consisted of 358 COVID-19, 8.066 normal and 5.538 pneumonia. Their CNN model was trained for 22 epochs and for 3 class classification achieved 93.30% accuracy. Similarly, in [35] the authors proposed a model called DarkCovidNet. In their tests, the dataset they used consisted of 127 COVID-19, 500 normal and 500 pneumonia. Their DL model was trained for 100 epochs and for 3 class classification achieved an accuracy of 87.02%.

Furthermore, in [36] compared five pre-trained CNN models (MobileNetV2, VGG19, Xception, InceptionV3 and InceptionResNetV2). They used two sets of data, the first set consisting of 224 COVID-19, 504 normal, 700 bacterial pneumonia, while the second set consisted of 224 COVID-19, 504 normal, 714 bacterial and viral pneumonia. Each CNN model was trained with the same hyperparameters for 10 epochs. The DL MobileNetV2 architecture in the first set for 3 class classification achieved 92.85% accuracy and in the second dataset achieved 94.72% accuracy. Misra et al. [37] proposed a multi-channel pre-trained ResNet architecture. The dataset they used consisted of 1.579 normal, 4.245 pneumonia and 184 COVID-19 images. Their DL model was trained up to 500 epochs with early stopping criteria. Their model for 3 class classification achieved 93.90% accuracy.

In [38] the authors evaluated the VGG16, VGG19, DenseNet201, InceptionResNetV2, InceptionV3, Resnet50 and MobileNetV2 models. The dataset they used in their tests consisted of 2.780 images of bacterial pneumonia, 231 of COVID-19 and 1.583 normal. Each CNN model was trained with the same hyperparameters for 300 epochs. The InceptionResNetV2 model for 3 class classification achieved 92.18% accuracy, while the MobileNetV2 85.47%. Narinet al. [39] compared five pre-trained CNN models (ResNet50, ResNet101, ResNet152, InceptionV3 and InceptionResNetV2). In their study they applied three different binary classifications with four classes (COVID-19, viral pneumonia, bacterial pneumonia, normal). The used dataset in their tests consisted of 341 COVID-19, 2.800 normal, 2.772 bacterial pneumonia and 1.493 viral pneumonia chest radiography images. Each DL model trained 30 epochs. The results showed that the pre-trained ResNet50 model provided the highest classification performance in all three different binary classifications with an accuracy of over 96%. The ResNet50 model they propose consists of over 25 million parameters.

Most of the above studies used a limited dataset, which varies and includes minimal samples of COVID-19 X-rays ranging from 100 samples up to 500 using millions of parameters. Studies have shown an urgent need for efficient DL models in large-scale data with fewer parameters that will provide higher accuracy. This work presents a modified version of the MobileNetV2 model for detecting COVID-19 infection from chest X-ray images and focuses on 4 class classifications.

In contrast with all other authors, we propose a well-tuned DL architecture that can be executed on a low-power embedded system in real time (sustainable performance with over 120 fps). Furthermore, we have trained our model using a dataset of over 21 k images. In our study, our goal is for the model to combine small network size a small number of parameters and to show very high performance for the classification of the four most important cases for lung deceases (COVID-19, normal, lung opacity and viral pneumonia) in a large dataset 21.165 chest X-ray images.

3. Materials and Methods

In this section, we will describe the dataset. Then, we will describe the details of the proposed model architecture, the training on the embedded GPU and the performance metrics.

3.1. Dataset Description

In this study, the dataset we used in our experiments is the COVID-19 Radiography Database [40]. This dataset is currently one of the largest public databases with 21.165 chest X-ray images and includes 3.616 COVID-19 positive cases, 10.192 normal, 6.012 lung opacity (Non-COVID lung infection) and 1.345 viral pneumonia images. All the images are in PNG file format, and the resolution is 299 × 299 pixels. Figure 1 shows a sample of the COVID-19 Radiography Database (normal, COVID-19, lung opacity and viral pneumonia).

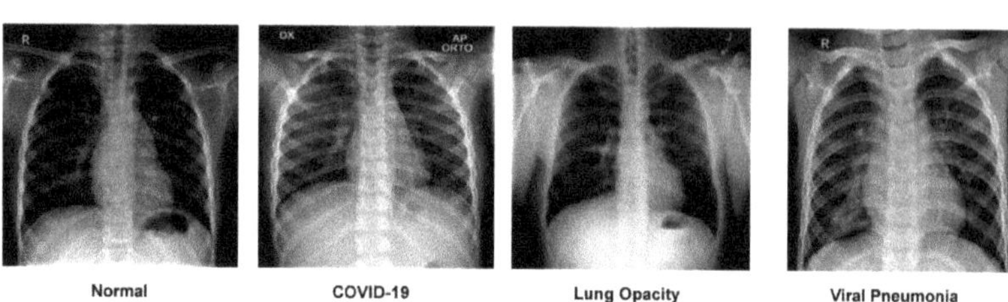

Figure 1. A sample of the COVID-19 Radiography Database.

3.2. Splitting the Dataset

We randomly split the dataset into training 70% (14.813 images), validation 20% (4.232 images), and testing 10% (2.120 images). The training set consists of 14.813 images, in which there were 2.530 COVID-19 images, 7.134 normal images, 4.208 lung opacity images and 941 viral pneumonia images. The validation set consists of 4.232 images, in which there were 723 COVID-19 images, 2.038 normal images, 1.202 lung opacity images and 269 viral pneumonia images. The test set consists of 2.120 images, in which there were 363 COVID-19 images, 1.020 normal images, 602 lung opacity images and 135 viral pneumonia images. The images used for testing were never used in the training process. The summary of the dataset (training, validation, and test) after the split is presented in Table 1.

Table 1. The COVID-19 Radiography Database summary used in our research.

Category	Train	Validation	Test
COVID-19	2530	723	363
Normal	7134	2038	1020
Lung Opacity	4208	1202	602
Viral Pneumonia	941	269	135
Total	14,813	4232	2120

3.3. Data Pre-Processing

Most CNNs are trained on image resolution with 224 × 224 pixels [41]. Therefore, in all the images of the COVID-19 Radiography Database [40] we changed the size from 299 × 299 pixels to 224 × 224 pixels. The size change was performed using the Python Imaging Library (PIL). The data augmentation methods such as random rotation, width shift, height shift, horizontal and vertical flip operations were not used because they have many limitations on medical images due to their strict format [42].

3.4. Proposed Modified Model Architecture

Most state-of-the-art CNNs have many millions of parameters to improve accuracy, and thus they are very time consuming during the training process. The proposed model is based on the MobileNetV2 architecture [41]. MobileNetV2 is one of the most popular lightweight network architectures for vision applications (classification, object detection and semantic segmentation). It is a very efficient model that improves the state-of-the-art performance on many visual recognition tasks. The MobileNetV2 is based on an inverted residual and linear bottleneck, significantly decreasing the number of operations and memory needed while retaining high accuracy. Although the classification accuracy of MobileNetV2 is 71% top-1 accuracy such as that of deep CNNs such as VGG16 and VGG19 (up 138 million parameters and 520 MB size) [27], it has the unique advantages of a smaller size network, fewer parameters, fewer operations, high efficiency, low-power and low latency.

The weights of the network are initialized with weights from a model pre-trained on ImageNet [43]. The top layer of the MobileNetV2 is discarded. To enhance the learning ability of the model, we added three additional new layers. In contrast with all other authors, we present an enhanced MobileNetV2 with more layers, improving the learning ability. The number and types of layers were determined after enough research into various networks and performing many different executions.

The first layer that we added is a global average pooling to minimize the overfitting and the number of parameters in the model by applying corresponding mathematical computation. Global average pooling generates one feature map for each corresponding category of the classification task in the last fully connected layer. The advantage of global average pooling is more native to the convolution structure between feature maps and categories [44].

After that, a fully connected layer with 512 nodes is added as a second layer with Rectified Linear Unit (ReLU) activation function [45]. The fully connected layer functions similar to a multilayer perceptron. Mathematical computation of ReLU activation function is shown in Equation (1).

$$ReLU(x) = \begin{cases} 0, & \text{if } x < 0 \\ x, & \text{if } x \geq 0 \end{cases} \tag{1}$$

A dropout layer is added, as the third layer, with a dropout rate of 20% to prevent network overfitting and divergence [46–48]. Dropout is a technique that randomly sets a portion of units, along with their connections, from the fully connected layers to zero during training. Dropout improves the performance of neural networks on supervised learning tasks in computer vision [48].

Finally, a fully connected layer is added as an output layer with 4 nodes, each one for every category, and SoftMax [49] is used as the activation function. This layer is used to predict output images. Mathematical computation of SoftMax activation function is shown in Equation (2).

$$Softmax(x_i) = \frac{e^{x_i}}{\sum_{y=1}^{m} e^{x_y}} \tag{2}$$

where x_i represent input data and m the number of classes.

The details of the architecture, parameters and output shape of the proposed model are presented in Table 2. It is important to notice that the global average pooling and the dropout layer consist of 0 parameters, while the fully connected layer with 512 nodes required our model to carry only 655.872 parameters but gave it higher accuracy than the standard architecture (omitting our 3 addition layers). Our model has a total of 2.915.908 parameters and is much lighter than many other studies that appear in the literature, which include more than 25 million parameters. The overall proposed methodology of the model is shown in Figure 2.

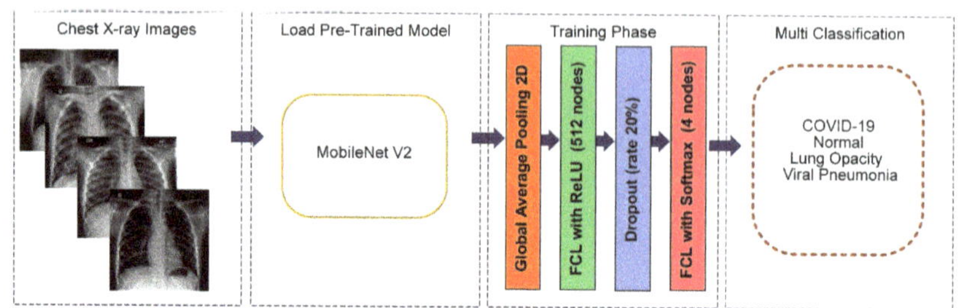

Figure 2. Modified MobileNetV2 architecture for multiclass classification problem (COVID-19, normal, viral pneumonia and lung opacity).

Table 2. Details of the proposed architecture, parameters and output shape.

Layer (Type)	Output Shape	Parameters
MobileNetV2 (Model)	$10 \times 10 \times 1290$	2,257,984
Global Average Pooling	0	0
Fully Connected Layer	512	655,872
Dropout	0	0
Fully Connected Layer (Classes)	4	2052
	Total Parameters: 2,915,908	

3.5. Training Strategy

We train our model using a two-step strategy. In the first step, we freeze the internal layers of the network and train only the three new layers (global average pooling layer, fully connected ReLU layer and dropout layer) for 10 epochs using an Adam optimizer [50] ($\beta1 = 0.9$ and $\beta2 = 0.999$) with a learning rate of 0.001. In this step, the learning is enhanced in the features of our dataset.

In the second step, we are reducing the learning rate of 0.0001 for 20 epochs and training throughout the network to inform all the weights in the network. The batch size of 32 remained constant throughout the training. In both phases of training, the activation function at intermediate layers is ReLU, and at the output layer is SoftMax. Loss function is categorical cross-entropy, and the dropout value is 0.20. The hyperparameters used are presented in Table 3.

3.6. Training the Model on the Embedded GPU

Training and testing are performed on the Nvidia Jetson AGX Xavier [51] with 512 CUDA cores, an 8 core ARM processor, 32 GB RAM, 32 GB eMMC and various other peripherals. Nvidia Jetson CPU-GPU heterogeneous architecture achieves high-performance computing and where can be quickly programmed to accelerate complex DL tasks [52]. The embedded GPU has shown great acceleration potential and involves low-power, high accuracy and efficiency in point-of-care medical applications [28,29,53,54]. At the same time, they provide local processing, eliminating security and privacy issues where required in biomedical systems [55]. Finally, we need a portable tool to facilitate medical diagnosis with low-weight, low-cost devices with accuracy, speed and power efficiency. The CNN models have trained on the Keras [56] framework.

The system can process up to 115 images per second suitable for real-time processing with approximately 28.5 Watts of maximum power consumption, which is much lower than the typical 1 kWatt for a PC with a GPU that executes very large DL models and cannot reach our highly optimized accuracy.

Table 3. The hyperparameters were used for all the experiments on the Jetson AGX Xavier embedded system.

Parameters	Value
Optimizer	Adam ($\beta1 = 0.9$ and $\beta2 = 0.999$)
Learning Rate	0.001–0.0001
Batch Size	32
Dropout value	0.20
Loss Function	Categorical cross-entropy
Activation function at intermediate layers	ReLU
Activation function at output layer	SoftMax
Total Training Epochs	30

3.7. Model Evaluation Metrics on the Test Dataset

Four metrics (accuracy, precision, recall and f1-score) [57] were used to evaluate the performance of the standard MobileNetV2 and the modified MobileNetV2 architectures. The calculation types of the metrics are shown in Equations (3)–(6), where *TP, FN, FP, TN* represent the number of true positives, false negatives, false positives and true negatives.

$$\text{Accuracy} = \frac{TP + TN}{TP + FN + FP + TN} \times 100\% \tag{3}$$

$$\text{Precision} = \frac{TP}{TP + FP} \times 100\% \tag{4}$$

$$\text{Recall} = \frac{TP}{TP + FN} \times 100\% \tag{5}$$

$$\text{F1-Score} = \frac{2 \times (\text{Precision} \times \text{Recall})}{(\text{Precision} + \text{Recall})} \times 100\% \tag{6}$$

4. Experimental Results

All executions were performed on the Jetson AGX Xavier embedded system. The total training times of standard MobileNetV2 were 1 h 38 min 22 s, and for the modified MobileNetV2, it was 1 h 40 min 52 s. Table 4 shows the performance comparison of the standard MobileNetV2 and the modified MobileNetV2. The total accuracy is 95.80% in the modified MobileNetV2, and the standard MobileNetV2 is 90.47%. The support is the number of occurrences of each class in the testing dataset, as shown in the classification report.

In Figures 3 and 4 we show the training accuracy/loss and validation accuracy/loss for standard MobileNetV2 and modified MobileNetV2, respectively. Modified MobileNetV2 architecture achieved 95.80% testing accuracy, while MobileNetV2 standard 90.47% for 4 classes of the public COVID-19 Radiography Database. It is clear that the performance of our modified MobileNetV2 model is much better than the standard MobileNetV2. Additionally, the modified MobileNetV2 model is more stable, while the standard MobileNetV2 shows more oscillation. The confusion matrix of the standard MobileNetV2 and proposed method is shown in Figures 5 and 6. It is evident that our modified MobileNetV2 achieves much better results, classifying correctly many more images than the standard MobileNetV2.

Table 4. Comparison of classification report of standard MobileNetV2 and modified MobileNetV2 for 4 classes (COVID-19, normal, viral pneumonia and lung opacity) classification.

	Modified MobileNetV2			Standard MobileNetV2			
Class	Precision	Recall	F1-Score	Precision	Recall	F1-Score	Support
COVID	0.9888	0.9697	0.9791	0.8880	0.9174	0.9024	363
Lung Opacity	0.9416	0.9369	0.9392	0.8680	0.8953	0.8814	602
Normal	0.9527	0.9667	0.9596	0.9312	0.9029	0.9169	1020
Viral Pneumonia	0.9923	0.9556	0.9736	0.9259	0.9259	0.9259	135
Accuracy			0.9580			0.9047	2120
Macro avg	0.9688	0.9572	0.9629	0.9033	0.9104	0.9067	2120
Weighted avg	0.9582	0.9580	0.9581	0.9055	0.9047	0.9049	2120

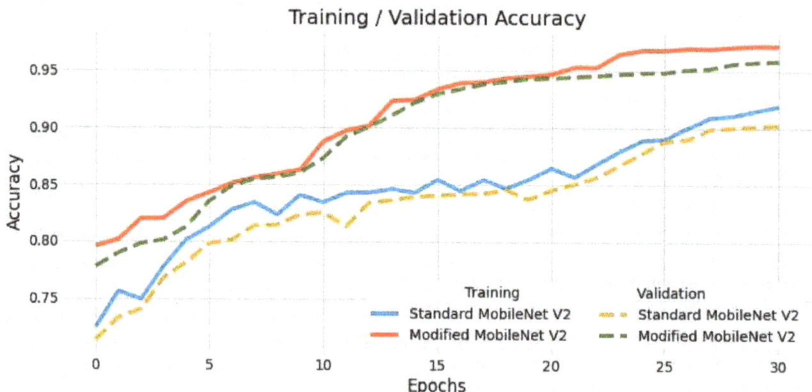

Figure 3. Comparison of training accuracy of standard MobileNetV2 and modified MobileNetV2. Our modified DL model presents a stable high accuracy due to its minimum loss per epoch.

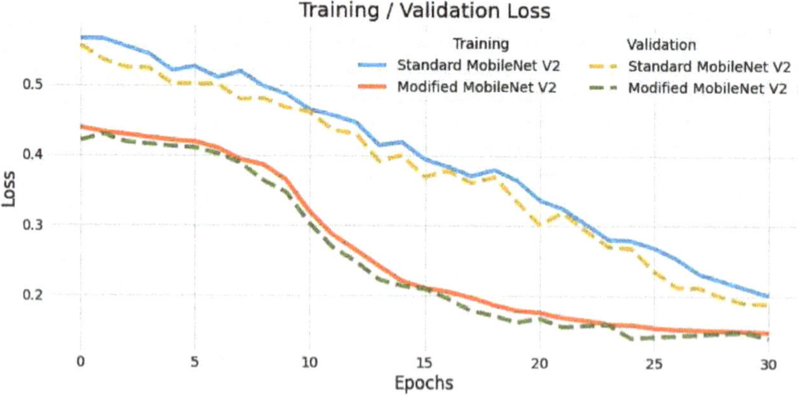

Figure 4. Comparison of training loss of standard MobileNetV2 and modified MobileNetV2. Our modified DL model has much lower loss in every epoch, illustrating its high optimized structure.

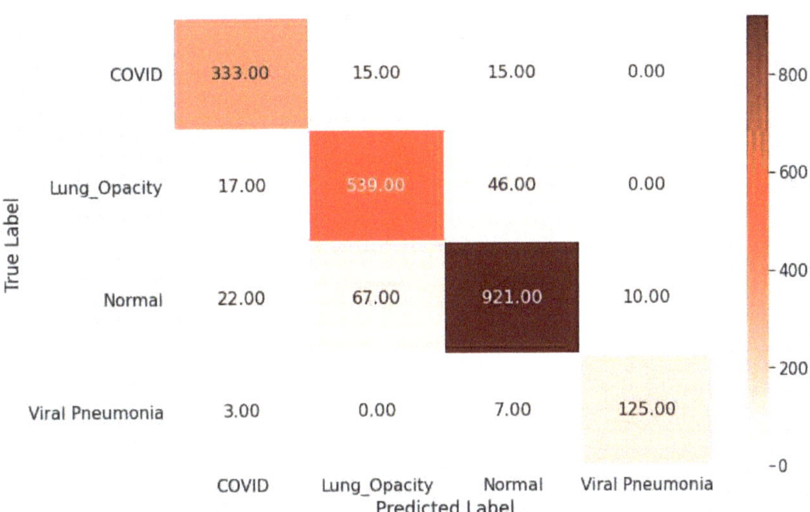

Figure 5. The confusion matrix of the standard MobileNetV2 shows many false negative and false positive classifications, especially for the COVID-19 cases.

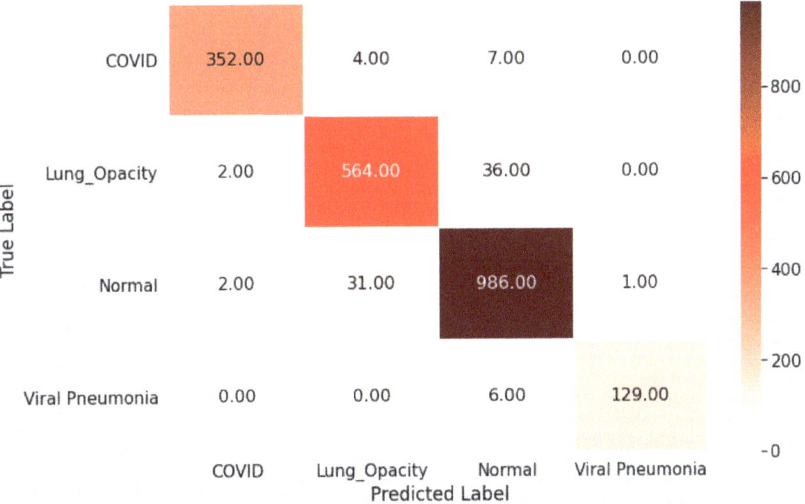

Figure 6. The confusion matrix of the proposed method (Modified MobileNetV2) shows that our optimized model predicts correctly in most cases much better than the standard DL model.

Comparison with Other Approaches

Comparing the performance of the proposed architecture with other existing methods is not possible because the dataset is different. Most studies are mainly concerned with binary classification, while there is a relatively limited number of studies concerning multiple classes.

However, in Table 5, we summarize the overall accuracy and architecture performance compared to our modified model from other existing models. Specifically, we only look at architectures used to classify 3 or 4 classes, but most have a small number of COVID-19 samples. It is evident that our model has much higher accuracy than every other research, backed up by the biggest dataset of X-rays examined.

As shown in Table 5, the proposed dataset presents the largest number of COVID-19 cases. We introduce a well-tuned DL technique that can be executed on a low-power embedded system in real time with promising results. We propose three additional new layers (global average pooling, fully connected layer with 512 nodes and dropout) and achieve high performance, small network size and a small number of parameters. We achieve the highest accuracy compared to the literature review with an accuracy of 95.80% for multiclass classification.

Table 5. Comparison of the proposed model with other existing models for COVID-19 classification using chest X-ray images for 3 or 4 class classification.

Study	Architecture	Dataset	Number of Classes	Number of Parameters (million)	Overall Accuracy (%)
[30]	MobileNetV2	186 COVID-19, 186 normal, 186 bacterial pneumonia and 186 viral pneumonia	4	3.5	95.40
[31]	Bayesian CNN	68 COVID-19, 1.583 normal, 2.786 bacterial pneumonia and 1.504 viral pneumonia	4	-	89.82
[32]	CoroNet	284 COVID-19, 310 normal, 330 bacterial pneumonia and 327 viral pneumonia	4	33	89.60
[33]	CovXNet	305 COVID-19, 305 normal, 305 viral pneumonia and 305 bacterial pneumonia	4	-	90.20
[34]	COVID Net	358 COVID-19, 8.066 normal and 5.538 pneumonia	3	11.75	93.30
[35]	DarkCovidNet	127 COVID-19, 500 normal and 500 pneumonia	3	3.1	87.02
[36]	MobileNetV2	224 COVID-19, 504 normal, 714 bacterial and viral pneumonia	3	-	94.72
[37]	ResNet	184 COVID-19, 1.579 normal and 4.245 pneumonia	3	25	93.90
[38]	MobileNetV2	231 COVID-19, 1.583 normal and 2.780 bacterial pneumonia	3	-	85.47
Proposed	Modified MobileNetV2	3.616 COVID-19, 10.192 normal, 6.012 lung opacity and 1.345 viral pneumonia	4	2.9	95.80

5. Conclusions and Future Work

COVID-19 has spread rapidly worldwide and is a severe public health problem. Thus, an early and valid prognosis of COVID-19 infected patients is vital to preventing the spread of the disease. This study proposes a lightweight neural network that provides high prediction accuracy from chest X-rays. The results of this study seem promising, but they must be further improved as more and more COVID-19 chest X-ray data become available. In fact, the model benefits of the numerosity of the examined X-ray dataset in reaching much higher accuracy than every other compared research. Our architecture approach achieves 95.80% accuracy, the highest up to this date compared with similar research, and processes up to 115 images per second with approximately 28.5 Watts of maximum power consumption on the Nvidia Jetson AGX Xavier. Furthermore, our work is the first to propose an optimized model for a low-power embedded system with increased detection accuracy. In a later study, we will test the performance of the proposed approach on different modern CNN models with an even larger number of COVID-19 and the number of classes chest X-rays in the dataset.

Author Contributions: Conceptualization, A.S., T.S. and D.T.; Formal analysis, A.S. and D.T.; Investigation, T.S. and A.S.; Methodology, T.S., D.T. and A.S.; Validation, A.S. and T.S.; Project administration, A.S. and M.D.; Supervision, M.D.; Writing—original draft, T.S., D.T. and A.S.; Writing—review and editing, A.S. and M.D. All authors have read and agreed to the published version of the manuscript.

Funding: This research received no external funding.

Institutional Review Board Statement: Not applicable

Informed Consent Statement: Not applicable

Data Availability Statement: Publicly available datasets were analyzed in this study. This data can be found here: https://www.kaggle.com/tawsifurrahman/covid19-radiography-database/activity (accessed on 25 May 2021).

Conflicts of Interest: The authors declare no conflict of interest.

Abbreviations

The following abbreviations are used in this manuscript:

AI	Artificial Intelligence
ARDS	Acute Respiratory Distress Syndrome
ARM	Architecture Reference Manual
CNN	Convolutional Neural Network
CPU	Central Processing Unit
CUDA	Compute Unified Device Architecture
DL	Deep Learning
eMMC	Embedded MultiMedia Card
FN	False Negative
FP	False Positive
GPU	Graphics Processing Unit
RAM	Random access memory
ReLU	Rectified Linear Unit
PC	Personal Computer
PIL	Python Imaging Library
PNG	Portable Network Graphics
TN	True Negative
TP	True Positive
WHO	World Health Organization

References

1. World Health Organization. Coronavirus Disease (COVID-19) Pandemic. Available online: https://www.who.int/ (accessed on 5 January 2022).
2. Worldometer. COVID-19 Coronavirus Pandemic. Available online: https://www.worldometers.info/coronavirus/ (accessed on 15 January 2022).
3. Huang, C.; Wang, Y.; Li, X.; Ren, L.; Zhao, J.; Hu, Y.; Zhang, L.; Fan, G.; Xu, J.; Gu, X.; et al. Clinical features of patients infected with 2019 novel coronavirus in Wuhan, China. *Lancet* **2020**, *395*, 497–506. [CrossRef]
4. Wang, D.; Hu, B.; Hu, C.; Zhu, F.; Liu, X.; Zhang, J.; Wang, B.; Xiang, H.; Cheng, Z.; Xiong, Y.; et al. Clinical characteristics of 138 hospitalized patients with 2019 novel coronavirus–infected pneumonia in Wuhan, China. *JAMA* **2020**, *323*, 1061–1069. [CrossRef]
5. Puderbach, M.; Eichinger, M.; Haeselbarth, J.; Ley, S.; Kopp-Schneider, A.; Tuengerthal, S.; Schmaehl, A.; Fink, C.; Plathow, C.; Wiebel, M.; et al. Assessment of morphological MRI for pulmonary changes in cystic fibrosis (CF) patients: Comparison to thin-section CT and chest X-ray. *Investig. Radiol.* **2007**, *42*, 715–724. [CrossRef]
6. Jaiswal, A.K.; Tiwari, P.; Kumar, S.; Gupta, D.; Khanna, A.; Rodrigues, J.J. Identifying pneumonia in chest X-rays: A deep learning approach. *Measurement* **2019**, *145*, 511–518. [CrossRef]
7. Bai, H.X.; Hsieh, B.; Xiong, Z.; Halsey, K.; Choi, J.W.; Tran, T.M.L.; Pan, I.; Shi, L.B.; Wang, D.C.; Mei, J.; et al. Performance of radiologists in differentiating COVID-19 from non-COVID-19 viral pneumonia at chest CT. *Radiology* **2020**, *296*, E46–E54. [CrossRef]
8. Chung, M.; Bernheim, A.; Mei, X.; Zhang, N.; Huang, M.; Zeng, X.; Cui, J.; Xu, W.; Yang, Y.; Fayad, Z.A.; et al. CT imaging features of 2019 novel coronavirus (2019-nCoV). *Radiology* **2020**, *295*, 202–207. [CrossRef]
9. Toussie, D.; Voutsinas, N.; Finkelstein, M.; Cedillo, M.A.; Manna, S.; Maron, S.Z.; Jacobi, A.; Chung, M.; Bernheim, A.; Eber, C.; et al. Clinical and chest radiography features determine patient outcomes in young and middle-aged adults with COVID-19. *Radiology* **2020**, *297*, E197–E206. [CrossRef]
10. Gattinoni, L.; Chiumello, D.; Caironi, P.; Busana, M.; Romitti, F.; Brazzi, L.; Camporota, L. COVID-19 pneumonia: Different respiratory treatments for different phenotypes? *Intensive Care Med.* **2020**, *46*, 1099–1102. [CrossRef]
11. Zech, J.R.; Badgeley, M.A.; Liu, M.; Costa, A.B.; Titano, J.J.; Oermann, E.K. Variable generalization performance of a deep learning model to detect pneumonia in chest radiographs: A cross-sectional study. *PLoS Med.* **2018**, *15*, e1002683. [CrossRef]

12. Hurt, B.; Kligerman, S.; Hsiao, A. Deep learning localization of pneumonia: 2019 coronavirus (COVID-19) outbreak. *J. Thorac. Imaging* **2020**, *35*, W87–W89. [CrossRef]
13. Longoni, C.; Bonezzi, A.; Morewedge, C.K. Resistance to medical artificial intelligence. *J. Consum. Res.* **2019**, *46*, 629–650. [CrossRef]
14. Sanida, T.; Varlamis, I. Application of Affinity Analysis Techniques on Diagnosis and Prescription Data. In Proceedings of the 2017 IEEE 30th International Symposium on Computer-Based Medical Systems (CBMS), Thessaloniki, Greece, 22–24 June 2017; pp. 403–408. [CrossRef]
15. Sanchez-Reyes, L.M.; Rodriguez-Resendiz, J.; Salazar-Colores, S.; Avecilla-Ramírez, G.N.; Pérez-Soto, G.I. A High-accuracy mathematical morphology and multilayer perceptron-based approach for melanoma detection. *Appl. Sci.* **2020**, *10*, 1098. [CrossRef]
16. Toledo-Perez, D.; Rodríguez-Reséndiz, J.; Gómez-Loenzo, R.A. A study of computing zero crossing methods and an improved proposal for EMG signals. *IEEE Access* **2020**, *8*, 8783–8790. [CrossRef]
17. Ortiz-Echeverri, C.J.; Salazar-Colores, S.; Rodríguez-Reséndiz, J.; Gómez-Loenzo, R.A. A new approach for motor imagery classification based on sorted blind source separation, continuous wavelet transform, and convolutional neural network. *Sensors* **2019**, *19*, 4541. [CrossRef]
18. Sánchez-Reyes, L.M.; Rodríguez-Reséndiz, J.; Avecilla-Ramírez, G.N.; García-Gomar, M.L.; Robles-Ocampo, J.B. Impact of eeg parameters detecting dementia diseases: A systematic review. *IEEE Access* **2021**, *9*, 78060–78074. [CrossRef]
19. Kermany, D.S.; Goldbaum, M.; Cai, W.; Valentim, C.C.; Liang, H.; Baxter, S.L.; McKeown, A.; Yang, G.; Wu, X.; Yan, F.; et al. Identifying medical diagnoses and treatable diseases by image-based deep learning. *Cell* **2018**, *172*, 1122–1131. [CrossRef]
20. Rajpurkar, P.; Irvin, J.; Zhu, K.; Yang, B.; Mehta, H.; Duan, T.; Ding, D.; Bagul, A.; Langlotz, C.; Shpanskaya, K.; et al. Chexnet: Radiologist-level pneumonia detection on chest x-rays with deep learning. *arXiv* **2017**, arXiv:1711.05225.
21. Shi, F.; Wang, J.; Shi, J.; Wu, Z.; Wang, Q.; Tang, Z.; He, K.; Shi, Y.; Shen, D. Review of artificial intelligence techniques in imaging data acquisition, segmentation and diagnosis for COVID-19. *IEEE Rev. Biomed. Eng.* **2020**, *14*, 4–15. [CrossRef]
22. Bullock, J.; Luccioni, A.; Pham, K.H.; Lam, C.S.N.; Luengo-Oroz, M. Mapping the landscape of artificial intelligence applications against COVID-19. *J. Artif. Intell. Res.* **2020**, *69*, 807–845. [CrossRef]
23. Tayarani-N, M.H. Applications of artificial intelligence in battling against COVID-19: A literature review. *Chaos Solitons Fractals* **2020**, *142*, 110338. [CrossRef]
24. Yang, W.; Sirajuddin, A.; Zhang, X.; Liu, G.; Teng, Z.; Zhao, S.; Lu, M. The role of imaging in 2019 novel coronavirus pneumonia (COVID-19). *Eur. Radiol.* **2020**, *30*, 4874–4882. [CrossRef]
25. Greenspan, H.; Van Ginneken, B.; Summers, R.M. Guest editorial deep learning in medical imaging: Overview and future promise of an exciting new technique. *IEEE Trans. Med. Imaging* **2016**, *35*, 1153–1159. [CrossRef]
26. Sanida, T.; Tsiktsiris, D.; Sideris, A.; Dasygenis, M. A Heterogeneous Lightweight Network for Plant Disease Classification. In Proceedings of the 2021 10th International Conference on Modern Circuits and Systems Technologies (MOCAST), Thessaloniki, Greece, 5–7 July 2021; pp. 1–4.
27. Simonyan, K.; Zisserman, A. Very deep convolutional networks for large-scale image recognition. *arXiv* **2014**, arXiv:1409.1556.
28. Paluru, N.; Dayal, A.; Jenssen, H.B.; Sakinis, T.; Cenkeramaddi, L.R.; Prakash, J.; Yalavarthy, P.K. Anam-Net: Anamorphic depth embedding-based lightweight CNN for segmentation of anomalies in COVID-19 chest CT images. *IEEE Trans. Neural Netw. Learn. Syst.* **2021**, *32*, 932–946. [CrossRef]
29. An, L.; Peng, K.; Yang, X.; Huang, P.; Luo, Y.; Feng, P.; Wei, B. E-TBNet: Light Deep Neural Network for Automatic Detection of Tuberculosis with X-ray DR Imaging. *Sensors* **2022**, *22*, 821. [CrossRef]
30. Kamal, K.; Yin, Z.; Wu, M.; Wu, Z. Evaluation of deep learning-based approaches for COVID-19 classification based on chest X-ray images. *Signal Image Video Process.* **2021**, *15*, 959–966.
31. Ghoshal, B.; Tucker, A. Estimating uncertainty and interpretability in deep learning for coronavirus (COVID-19) detection. *arXiv* **2020**, arXiv:2003.10769.
32. Khan, A.I.; Shah, J.L.; Bhat, M.M. CoroNet: A deep neural network for detection and diagnosis of COVID-19 from chest x-ray images. *Comput. Methods Programs Biomed.* **2020**, *196*, 105581. [CrossRef]
33. Mahmud, T.; Rahman, M.A.; Fattah, S.A. CovXNet: A multi-dilation convolutional neural network for automatic COVID-19 and other pneumonia detection from chest X-ray images with transferable multi-receptive feature optimization. *Comput. Biol. Med.* **2020**, *122*, 103869. [CrossRef]
34. Wang, L.; Lin, Z.Q.; Wong, A. COVID-net: A tailored deep convolutional neural network design for detection of COVID-19 cases from chest x-ray images. *Sci. Rep.* **2020**, *10*, 19549. [CrossRef]
35. Ozturk, T.; Talo, M.; Yildirim, E.A.; Baloglu, U.B.; Yildirim, O.; Acharya, U.R. Automated detection of COVID-19 cases using deep neural networks with X-ray images. *Comput. Biol. Med.* **2020**, *121*, 103792. [CrossRef] [PubMed]
36. Apostolopoulos, I.D.; Mpesiana, T.A. COVID-19: Automatic detection from x-ray images utilizing transfer learning with convolutional neural networks. *Phys. Eng. Sci. Med.* **2020**, *43*, 635–640. [CrossRef] [PubMed]
37. Misra, S.; Jeon, S.; Lee, S.; Managuli, R.; Jang, I.S.; Kim, C. Multi-channel transfer learning of chest x-ray images for screening of COVID-19. *Electronics* **2020**, *9*, 1388. [CrossRef]
38. El Asnaoui, K.; Chawki, Y. Using X-ray images and deep learning for automated detection of coronavirus disease. *J. Biomol. Struct. Dyn.* **2020**, *39*, 3615–3626. [CrossRef] [PubMed]

39. Narin, A.; Kaya, C.; Pamuk, Z. Automatic detection of coronavirus disease (COVID-19) using x-ray images and deep convolutional neural networks. *Pattern Anal. Appl.* **2021**, *24*, 1207–1220. [CrossRef]
40. Kaggle. COVID-19 Radiography Dataset. Available online: https://www.kaggle.com/tawsifurrahman/covid19-radiography-database/activity (accessed on 25 May 2021).
41. Sandler, M.; Howard, A.; Zhu, M.; Zhmoginov, A.; Chen, L.C. Mobilenetv2: Inverted residuals and linear bottlenecks. In Proceedings of the IEEE Conference on Computer Vision and Pattern Recognition, Salt Lake City, UT, USA, 18–23 June 2018; pp. 4510–4520.
42. Yadav, S.S.; Jadhav, S.M. Deep convolutional neural network based medical image classification for disease diagnosis. *J. Big Data* **2019**, *6*, 113. [CrossRef]
43. Deng, J.; Dong, W.; Socher, R.; Li, L.J.; Li, K.; Fei-Fei, L. Imagenet: A large-scale hierarchical image database. In Proceedings of the 2009 IEEE Conference on Computer Vision and Pattern Recognition, Miami, FL, USA, 20–25 June 2009; pp. 248–255.
44. Lin, M.; Chen, Q.; Yan, S. Network in network. *arXiv* **2013**, arXiv:1312.4400.
45. Hinton, G.E. Rectified linear units improve restricted boltzmann machines vinod nair. *Citeseer* **2010**, *7*, 1–8.
46. Hinton, G.E.; Srivastava, N.; Krizhevsky, A.; Sutskever, I.; Salakhutdinov, R.R. Improving neural networks by preventing co-adaptation of feature detectors. *arXiv* **2012**, arXiv:1207.0580.
47. Hawkins, D.M. The problem of overfitting. *J. Chem. Inf. Comput. Sci.* **2004**, *44*, 1–12. [CrossRef]
48. Srivastava, N.; Hinton, G.; Krizhevsky, A.; Sutskever, I.; Salakhutdinov, R. Dropout: A simple way to prevent neural networks from overfitting. *J. Mach. Learn. Res.* **2014**, *15*, 1929–1958.
49. Goodfellow, I.; Bengio, Y.; Courville, A. *Deep Learning*; MIT Press: Cambridge, MA, USA, 2016.
50. Kingma, D.P.; Ba, J. Adam: A method for stochastic optimization. *arXiv* **2014**, arXiv:1412.6980.
51. Nvidia Developer. Jetson AGX Xavier Developer Kit. Available online: https://developer.nvidia.com/embedded/jetson-agx-xavier-developer-kit (accessed on 30 May 2021).
52. Mittal, S.; Vetter, J.S. A survey of CPU-GPU heterogeneous computing techniques. *ACM Comput. Surv. (CSUR)* **2015**, *47*, 1–35. [CrossRef]
53. Ardiyanto, I.; Nugroho, H.A.; Buana, R.L.B. Deep learning-based diabetic retinopathy assessment on embedded system. In Proceedings of the 2017 39th Annual International Conference of the IEEE Engineering in Medicine and Biology Society (EMBC), Jeju, Korea, 11–15 July 2017; pp. 1760–1763.
54. Page, A.; Shea, C.; Mohsenin, T. Wearable seizure detection using convolutional neural networks with transfer learning. In Proceedings of the 2016 IEEE International Symposium on Circuits and Systems (ISCAS), Montreal, QC, Canada, 22–25 May 2016; pp. 1086–1089.
55. Attaran, N.; Puranik, A.; Brooks, J.; Mohsenin, T. Embedded low-power processor for personalized stress detection. *IEEE Trans. Circuits Syst. II Express Briefs* **2018**, *65*, 2032–2036. [CrossRef]
56. Géron, A. *Hands-on Machine Learning with Scikit-Learn, Keras, and TensorFlow: Concepts, Tools, and Techniques to Build Intelligent Systems*; O'Reilly Media: Sebastopol, CA, USA, 2019.
57. Hossin, M.; Sulaiman, M. A review on evaluation metrics for data classification evaluations. *Int. J. Data Min. Knowl. Manag. Process.* **2015**, *5*, 1.

 *technologies*

MDPI

Article

Effective Current Pre-Amplifiers for Visible Light Communication (VLC) Receivers

Simon-Ilias Poulis [1,*], Georgios Papatheodorou [1], Christoforos Papaioannou [1], Yiorgos Sfikas [1], Marina E. Plissiti [1], Aristides Efthymiou [1], John Liaperdos [2] and Yiorgos Tsiatouhas [1]

1 Department of Computer Science & Engineering, University of Ioannina, GR-45110 Ioannina, Greece; g.papatheodorou@uoi.gr (G.P.); chpapaioannou@cs.uoi.gr (C.P.); gsfikas@uoi.gr (Y.S.); marina@uoi.gr (M.E.P.); efthym@uoi.gr (A.E.); tsiatouhas@cse.uoi.gr (Y.T.)
2 Department of Digital Systems, School of Economy and Technology, University of the Peloponnese, GR-23100 Sparta, Greece; i.liaperdos@uop.gr
* Correspondence: s.poulis@uoi.gr

Abstract: Visible light communication (VLC) is an upcoming wireless communication technology. In a VLC system, signal integrity under low illumination intensity and high transmission frequencies are of great importance. Towards this direction, the performance of the analog front end (AFE) sub-system either at the side of the transmitter or the receiver is crucial. However, little research on the AFE of the receiver is reported in the open literature. Aiming to enhance signal integrity, three pre-amplification topologies for the VLC receiver AFE are presented and compared in this paper. All three use bipolar transistors (BJT): the first consists of a single BJT, the second of a double BJT in cascade connection, and the third of a double BJT in Darlington-like connection. In order to validate the performance characteristics of the three topologies, simulation results are provided with respect to the light illumination intensity, the data transmission frequency and the power consumption. According to these simulations, the third topology is characterized by higher data transmission frequencies, lower illuminance intensity and lower power consumption per MHz of operation.

Keywords: visible light communication (VLC); wireless communication; analog front end (AFE); pre-amplification

Citation: Poulis, S.-I.; Papatheodorou, G.; Papaioannou, C.; Sfikas, Y.; Plissiti, M.E.; Efthymiou, A.; Liaperdos, J.; Tsiatouhas, Y. Effective Current Pre-Amplifiers for Visible Light Communication (VLC) Receivers. *Technologies* **2022**, *10*, 36. https://doi.org/10.3390/technologies10010036

Academic Editors: Spiros Nikolaidis and Rodrigo Picos

Received: 31 December 2021
Accepted: 11 February 2022
Published: 21 February 2022

Publisher's Note: MDPI stays neutral with regard to jurisdictional claims in published maps and institutional affiliations.

1. Introduction

Visible light communication (VLC) is a promising scientific field in which the interest of the research community has increased significantly during the last decade. The term VLC stands for a wireless communication system which uses the visible spectrum of light (380 nm to 750 nm, 430 THz to 750 THz) as the medium for transmitting information. Although the idea of using visible light in communications is not new [1], it was not until the beginning of the year 2000 that there were the first VLC experiments using light-emitting diode (LED) lamps. In fact, in [2] a white LED was used for simultaneous illumination and data transmission in an indoor space. That research was a springboard for optical wireless communications (OWC), increasing the research interest and leading to major innovations, such as new configuration techniques, new LED technologies and more. The year 2011 was a milestone for VLC. Initially, a first demonstration of a Light Fidelity (Li-Fi) network during a TEDx talk by Harald Haas [3] highlighted the innovation brought about by this promising wireless communication and resulted in an exponential increase on research output related to VLC, OWC and free-space optical communications (FSOC). Later in the same year, the IEEE included OWC/VLC in the official IEEE Standard for local and metropolitan area networks [4].

There are different types of OWCs that can use LEDs or laser diodes (LDs) as transmitters and photodiodes (PDs) or image sensors (IS) as receivers. Also, depending on the proposed technology, the physical means of communication may be the infrared (IR), visible light (VL) [5,6] or ultraviolet (UV) part of the electromagnetic radiation of light [7]. This new technology has some significant advantages such as the use of existing lighting infrastructure both for illumination and communication. Considering the gradual replacement of traditional lighting by LED lamps, the energy consumed for lighting is useful, in a twofold manner. However, in mobile devices the energy consumption by the LEDs used remains considerable.

The bandwidth of visible light is one of VLC's most important advantages over radio frequency (RF) communications. It is about 1000 times wider than the spectrum of radio frequencies, thus significantly reducing the possibility of channel congestion. In addition, the high frequency of electromagnetic waves of visible light (of the order of THz) can allow very high-speed data transmission. More specifically, compared to Wi-Fi, where the highest data rate reaches 1 Gbps (in the Wireless Gigabit (WiGig) standard [8]), studies on optical communications report data rates up to 100 Gbps, as in [9] where a LD is used.

VLC can also provide significant advantages in data security options compared to radio wave links, since visible light waves do not have the ability to penetrate through opaque solid surfaces like radio waves, making VLC safer than radio wave communications. In addition, all electromagnetic radiation that does not belong to the visible spectrum, even when very close to it, such as infrared and ultraviolet, tend to be harmful to human health or cause interference to machines, such as pacemakers and other important medical equipment. On the contrary, radiation belonging to the visible spectrum does not cause such interference and is harmless, making VLC safe for humans.

Apart from the above, VLC technology demonstrates a wide range of applications, such as internet connectivity, underwater communications, and even interplanetary communications. Furthermore, there are three applications that mainly attract research interest: (i) indoor VLC systems, (ii) transport and vehicular VLC systems and (iii) indoor positioning systems.

The study of indoor VLC systems arose due to the long periods that people spend inside buildings and the idea started to become a reality due to the gradual prevalence of LED lamps. In [10] the authors studied the possibility of using LEDs for VLC systems, presenting the requirements for such a system to act for lighting and communication at the same time. Nowadays, numerous buildings provide lighting with multiple LEDs where multiple input multiple output (MIMO) techniques can be applied, regardless of the type of physical medium configuration. In [11] a MIMO orthogonal frequency division multiplexing (MIMO–OFDM) system with a transmission speed of 220 Mbps at a distance up to 1 m between the transmitter and the receiver is reported, while in a recent work of the same research team [12], a speed of 1.1 Gbps was achieved at the same conditions.

In a VLC system, data transmission takes place, even when the light in an indoor space is dim or even switched off. In [13] the authors proposed a VLC system in which data transfer is performed even if the lamp remains switched off for human perception or at low intensity.

Using VLC in vehicles is also very advantageous, since its application cost can be relatively low, as the use of LED lamps is trivial. Also, a variety of light sources is available on the roads, such as car headlights or traffic lights, which can be used to develop intelligent transport systems (ITS) [14], as they can establish a line of sight (LoS) VLC system. A vehicular VLC (V2LC) system requires one or more mobile (vehicles) and fixed (traffic lights, streetlights) nodes. The nodes need to obtain transmitters and receivers to create a dynamic communication network, to transmit any useful information collected by sensors embedded in the vehicles and the environment [15], while at the same time an important possibility for these networks is a direct vehicle-to-vehicle (V2V) communication.

With the development of the Internet of Things (IoT) and the continuous growth of mobile devices, new applications are available offering new possibilities to the end users. Such features are related to personal navigation and information or advertising depending on the user's location. In order to achieve the above features, location capabilities are required. VLC technology can provide indoor localization, with great precision using existing lighting–communication infrastructure. Thus, the receiver will capture the signals off the LEDs and with appropriate algorithms the exact position will be determined.

High-precision internal positioning systems, such as Epsilon [16] and Luxapose [17], have come to light since 2014. In the former, many LED lamps (transmitters) and photodiodes (receivers) were used to easily implement a low-cost system, with the ability to find the position with an error of 0.4 m. In Luxapose, the receivers were image sensors (IS) from a smartphone camera, having only a 0.1 m internal detection error, while providing the ability to orient the device.

VLC is therefore a promising communication technology, which will be of great concern to the research community in the coming years, either as a complementary technology to the already existing wireless communications, or as the main means of wireless communication. Although the pertinent IEEE standard has been revised since 2018 [18], research is still at an early stage in this field, as the full operating framework has not yet been defined.

As mentioned above, the range of applications for optical wireless communication is wide resulting in a dispersion of research towards various directions. Also, beyond the applications, the attention of the research community has focused on various parts of the physical layer (PHY), such as the transmitter–receiver architecture, the coding–signal modulation and the multiplexing of multiple users.

Since the initial release of the IEEE standard [4], different modulation and coding schemes have been proposed at the PHY level, depending on the application usage of a VLC system. Typical modulation schemes are on-off keying (OOK), color shift keying and variable pulse position modulation (VPPM) for VLC systems using photodetectors such as PDs. Respectively, in the revised version of the standard [18] modulation schemes were added such as camera on-off keying (C-OOK), rolling shutter frequency shift keying (RS-FSK) and others, related to the use of the IS of digital cameras.

In addition to the proposed standard schemes, the research interest has turned to various data multiplexing schemes such as OFDM, MIMO, wavelength division multiplexing (WDM) and others or combinations thereof, in order to achieve high transmission rates. For example, in [12] an optical modulation technique for MIMO–OFDM is proposed, while in [19] a performance improvement of M-quadrature amplitude modulation (M-QAM) OFDM–non-orthogonal multiple access (NOMA) is presented. At the same time, in [20] a 16-QAM OFDM transmission scheme with rates of 4 Gbit per second is proposed and in [21] a LED-based wavelength division multiplexing (WDM) scheme achieving a 10 Gbps data rate is presented. Moreover, an additional area of interest for a more complete coverage of the optical channel are the multiple access methods, with NOMA so far seeming to gain most research interest [22,23], due to the significant improvement of the spectral performance of the channel.

Also, an important issue that arises in VLC is the elimination of flickering using coding schemes, in the existing formatted optical signal, such as run-length limiting (RLL) and Reed–Solomon (RS) encoders, with simultaneous clock recovery and error detection capability [18]. Although the above encoding schemes are well defined by the standard, there is considerable research on alternative ways of error detection and correction, such as the use of polar code (PC) instead of RS in CSK modulation VLC systems [24] or with a protograph low-density parity-check (LDPC) code of two types [25].

Today, the research mainly focuses on the transmitter with respect to the power supply, the LED characteristics [26] and the LED driving sub-circuit [27,28]. However, considering the receiver of a VLC system, the research activity is not extensive enough. Meanwhile, the AFE of the receiver constitutes an equally important part of a VLC system. It receives the visual signal and converts it to an electrical one that is filtered and amplified in order to be digitized and processed. The limited number of studies in the literature are mainly focused on the receiver architecture at a basic level, such as the use of the trans-impedance amplifier (TIA) to amplify the received signal [29]. Therefore, the research on the AFE of the receiver is essential as it contributes to the smooth operation of a VLC system.

In this paper, three effective current pre-amplification topologies that supplement the AFE of VLC receivers are presented and simulation results are included. The three topologies are compared with respect to their characteristics that are related to, (1) the illuminance falling on the photosensitive surface of the receiver photodiode, (2) the transmission frequency of the optical signal and (3) the total power consumption of the receiver. The paper is organized as follows. In Section 2 preliminaries on VLC systems are given. Next, in Section 3 the AFE of a VLC system is discussed and the three current pre-amplification topologies are presented. In Section 3, simulation results are provided on the performance characteristics of the three topologies. Finally, in Section 4 the simulation results are discussed.

2. Preliminaries

In a VLC system, information is transmitted through the visible light. The transmitter modulates the light (e.g., on-off switching or continuous changes of light intensity) in a way that is not perceived by the human eye. In this way, both the illumination and the information transmission needs are covered simultaneously. The operation is achieved either by using LEDs that support both functions or by using dedicated LEDs for data transmission while supplementing the required room illumination with other light sources that are exploited only for lighting.

From the above discussion, it is well understood that the information transmission must be carried out at a frequency that the human eye cannot perceive. This frequency has been already defined by the pertinent IEEE standard at above 200 Hz [4] and reconfigured by its revised version in 2018 [18] at above 2 kHz.

On the receiver side, the optical signal can be retrieved primarily either by using a photodiode or the sensor of the built-in camera of a portable device. These two approaches have essential differences that are exploited by several applications depending on the data transmission frequency. For example, since a built-in camera supports lower data rates (in the order of a few kHz), it can be used in an indoor location and navigation system. On the contrary, as photodiodes can respond to much higher frequencies (in the range of a few MHz) their use in applications that support higher data rates is preferred.

Furthermore, the reverse current of a photodiode may be in the range of some tenths of μA up to 200 μA depending on the photodiode model. This feature can be crucial in a VLC system since such a small input current signal may distort data resulting in incorrect reception of the transmitted information.

3. Visible Light Communication (VLC) Receiver and Pre-Amplification Circuits

Our work focuses on a receiver circuit that uses a photodiode as the optical signal sensor. The basic structure of the AFE of the VLC receiver under consideration consists of: (a) a photodiode (PD), (b) a pre-amplification circuit (Pre-Amp), (c) a TIA, (d) a DC cancellation circuit (DCC), (e) a passive filter (PF) and (f) a voltage amplifier (VA), as shown in the block diagram of Figure 1.

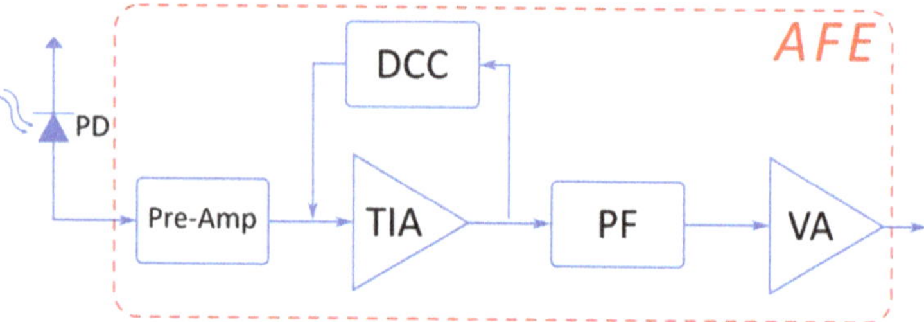

Figure 1. The analog front end (AFE) of the receiver under consideration.

More specifically, the optical signal is converted through the photodiode into a current signal. This signal, after its pre-amplification, drives the TIA, which is supported by the DCC circuit, for the conversion to a voltage signal without its DC offset. Then, the signal is passed through the passive filter that eliminates any unwanted frequencies. Finally, the signal is amplified at appropriate levels by the VA in order to be digitized and drive a microcontroller or microprocessor, which will decode it and retrieve the information.

3.1. VLC Receiver

In more detail, for the VLC receiver design the TIA–DCC sub-circuits proposed in [30] were adopted in our work, while the anti-aliasing filter used in [30] was replaced by a high-pass filter and a voltage amplifier, as shown in Figure 2.

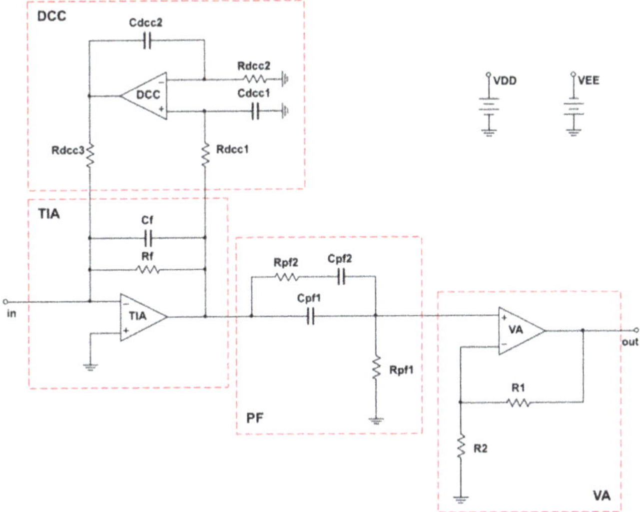

Figure 2. Detailed topology of the AFE circuitry after the pre-amplification block.

The combination of the TIA and DCC blocks ("Ambient Light Filter" in [30]) achieves the elimination of the DC offset from the signal and has the ability to self-adjust according to the average light intensity changes that are taking place in the lighting of the room. It is a very versatile active filter with its cut-off frequency set at 340 Hz.

The signal provided by the TIA–DCC blocks, without the DC offset, enters a high-pass passive filter with a cut-off frequency of 3 MHz. Finally, the filtered signal enters a voltage amplifier, powered by a single (positive) supply V_{CC} at 3.3 V, which acts as a half rectifier eliminating the negative part of the signal. The gain of the VA is 43.58 dB which is large enough to support easy digitization of the signal.

3.2. Pre-Amplification Circuit Topologies

According to our observations, the above typical receiver AFE topology presents serious signal integrity problems at low light intensities. For that reason, three pre-amplification topologies are explored in this work. The common feature of all three topologies is that bipolar transistors were chosen as their basic unit. Thus, the current of the photodiode controls the collector current of a transistor, forming a current source controlled by current.

The three topologies under consideration are: (i) a single BJT current amplifier, (ii) a double BJT in a cascade connection current amplifier and (iii) a double BJT in Darlington-like connection current amplifier. These topologies are presented in detail in the following paragraphs.

3.2.1. Single Bipolar Transistor (BJT)

The first pre-amplification topology, along with the photodiode, is shown in Figure 3. It is the simplest topology consisting of a bipolar transistor where its base is fed by the current signal produced by the photodiode. The base current is amplified at the collector according to the BJT current gain factor (h_{FE}) and the amplified current signal drives the TIA–DCC blocks.

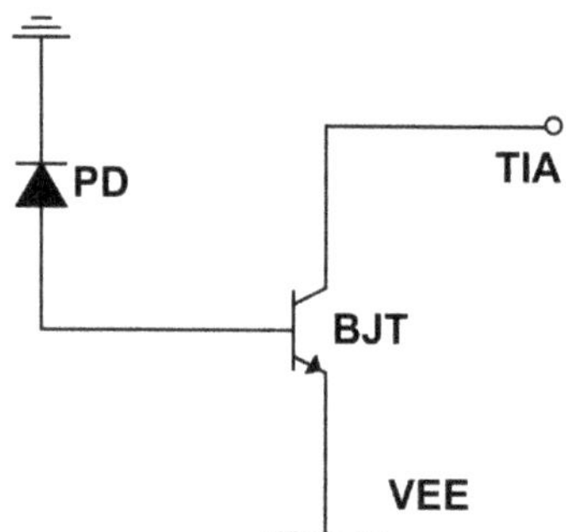

Figure 3. Single bipolar transistor (BJT) pre-amplification circuitry.

3.2.2. Double BJT in Cascade Connection

The second topology consists of two BJTs in cascade connection as shown in Figure 4. Its basic operation principle lies in the amplification provided by each BJT. Here, the incoming current signal is amplified via the BJT current gain factor h_{FE1} and the emitter current of the first BJT is obtained. The emitter current of BJT1 feeds the base of BJT2 which in turn is amplified by the current gain factor h_{FE2} and drives the TIA–DCC block. It should be noted that the RC network (with cutoff frequency of 80 kHz) is used to limit the DC offset of the current on the BJT1 collector and consequently is exploited to control the collector current of the first stage.

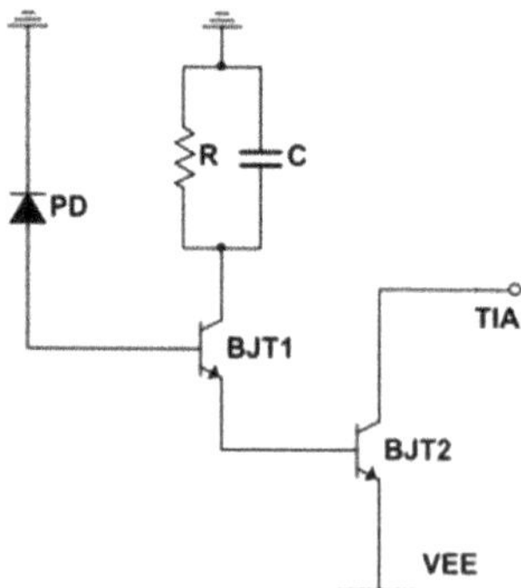

Figure 4. Double BJT in cascade connection for the pre-amplification circuitry.

3.2.3. Double BJT in Darlington-Like Connection

This topology also uses two BJTs but in a Darlington-like connection. Figure 5 presents the circuit which differs with respect to a typical Darlington scheme, due to the existence of the resistor at the collector of BJT1. This resistance is exploited to regulate the amplification and consequently to provide a better control of the amplified current signal that feeds the TIA-DCC block. As mentioned earlier, the BJT1 collector resistance provides a self-control capability of the current at the base of BJT2, aiming to reduce the DC current amplification.

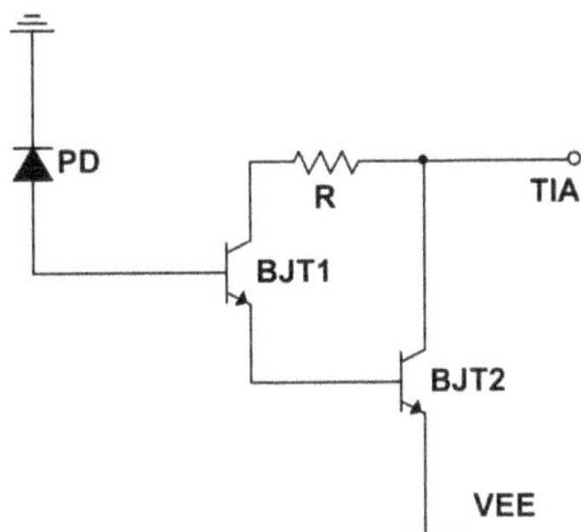

Figure 5. Double BJT in Darlington-like connection for the pre-amplification circuitry.

4. Simulation Results

In this section, initially, the components in Figures 2–5 that were used for the design of the VLC receiver under consideration are discussed. Next, the electrical model of the photodiode is presented and the structure of a typical signal that feeds a VLC receiver is analyzed. Both are exploited in the simulations of the receiver, which were performed in the LTSpice XVII platform of Analog Devices. Finally, the simulation results are presented that provide comparisons among the three pre-amplification topologies.

4.1. VLC Receiver Design

The AFE circuitry of the VLC receiver in Figure 2 along with the three pre-amplification topologies were designed using the schematic editor of the LT-SPICE platform. The components of the various parts in the design are presented in Table 1. For each one of them the Spice models of the manufacturers are exploded at the simulations, except the PD. The selection of the components was based on their ability to support the specifications of the receiver that are related to the response time, the bandwidth, the operating frequency and the slew rate.

Table 1. The components used in the AFE design of the receiver.

AFE Parts	Component
Photodiode (PD)	SFH–206K
BJT (in all topologies)	BFR–340F
Operational Amplifier used in TIA + VA	LTC6228
Operational Amplifier in DCC	ADA4622

The OSRAM SFH-206K [31] photodiode was chosen, as this component has a spectral sensitivity between 400–1100 nm, which is satisfactory for a VLC system. In addition, it has quite small rise and fall times, along with an appropriate current intensity and linearity with respect to the incident luminous flux per unit area of the photodiode. For the needs of our VLC system, the photodiode provides a 40μA current for an incident luminous power of 500 Lux on the surface of the photodiode.

The Analog Devices LTC6228 [32] low-distortion operational amplifier was selected for the TIA and the VA. This opamp is characterized by low noise, down to 0.88 nV/$\sqrt{\text{Hz}}$, high slew rate that reaches 500 V/us and the rail-to-rail output ability.

Unlike the cases of the TIA and VA, for the selection of the operational amplifier for the DCC sub-circuit there are no special specifications, as the DCC is the least demanding part of the receiver. Thus, the Analog Devices ADA4622 [33] opamp was chosen.

Finally, the Infineon Technologies BFR-340F [34] bipolar transistor was selected for the pre-amplification topologies under consideration. This component is an RF bipolar transistor with 14 GHz typical transition frequency and low capacitance coupling between its terminals.

4.2. Photodiode Electrical Model

To conduct the simulations of the receiver, an appropriate electrical model for the photodiode under consideration was necessary. Towards this direction, a proper sub-circuit was developed, according to the photodiode datasheet, using the typical topology shown in Figure 6 for a photodiode in reverse bias. In this model, the photodiode current is correlated with the actual illuminance intensity values.

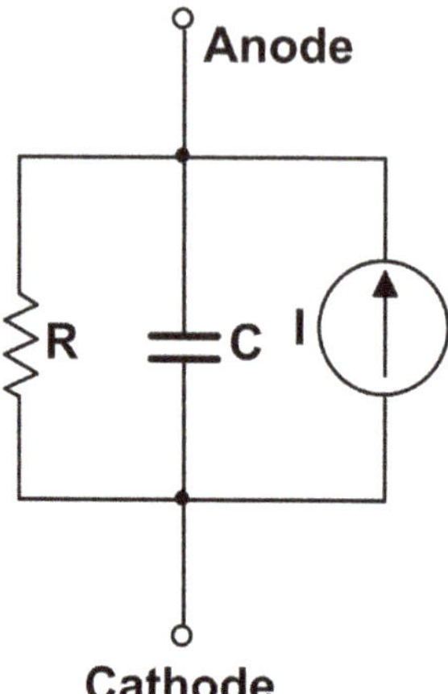

Figure 6. Typical topology of the electrical model of the photodiode.

The photodiode model consists of a current source, a capacitor and an ohmic resistor in parallel connection, as it is depicted in Figure 6. Capacitance and resistance values correspond to the SFH–206K datasheet values, while the current source amplitude is determined by the linear relationship between illuminance and photocurrent, as it is extracted from the pertinent datasheet diagram. Furthermore, the rise and fall times of the photodiode were also taken into account in this model.

In order to use a realistic current signal, it was necessary to consider an appropriate data-encoding method according to the IEEE standard [18]. The Manchester encoding was chosen, where the logic high bit ("1") is encoded with a low to high transition during one period of the system clock (Figure 7a), while the logic low bit ("0") is encoded with a high to low transition accordingly (Figure 7b).

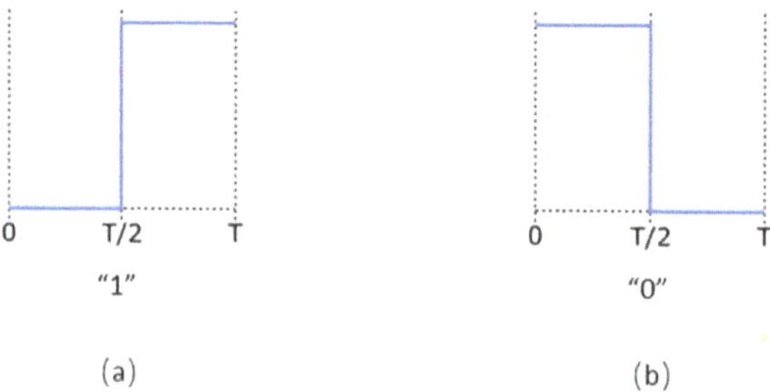

Figure 7. Manchester encoding (**a**) the logic high bit ("1") and (**b**) the logic low bit ("0").

Manchester encoding, although simple, is essential in a VLC system since it prevents an intense change in the brightness of the transmitter's light in the case of consecutive "1" or "0" bits. Thus, the human eye is protected from perceiving light brightness perturbations since it keeps a constant average light intensity, for signal frequencies above the minimum specifications of the standard.

In order to simulate a realistic operating scenario for the receiver, a Manchester-encoded 24-bit input sequence was created that is followed by an empty span of 16-clock periods. This results in the creation of a data package and a blank space after that, which is continuously repeated as shown in Figure 8. The generated sequence controls the current source of the photodiode model in Figure 6. The pulse amplitude and max frequency are determined according to the incident luminous power and the system clock frequency, respectively.

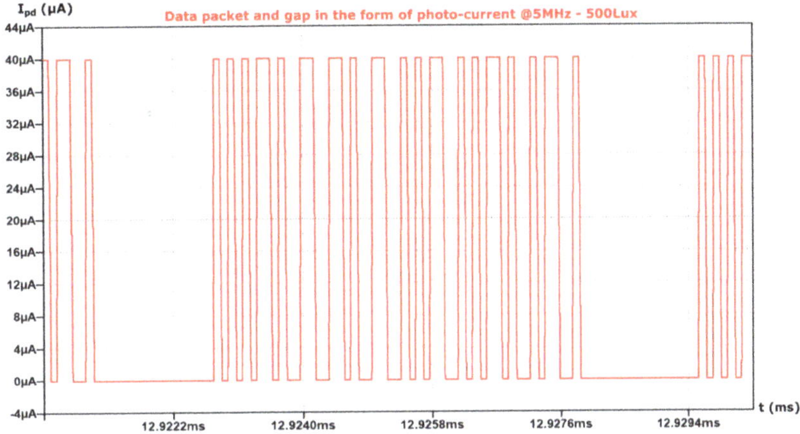

Figure 8. The data sequence that controls the photodiode model in the receiver simulations for an incident luminous power of 500 Lux and a frequency of 5 MHz.

Figure 8 illustrates the generated and "transmitted" 24-bit data sequence, according to the Manchester encoding. This data sequence (package) consists of pulses of longer or shorter duration. A short duration pulse corresponds to the transmission of the same bit value repeatedly, while a longer pulse corresponds to a bit change. According to Figure 8, in our simulations a package consists of 19 pulses, with a period equal to the system clock period (T) for a short pulse and a period 2T for a long pulse.

In practice, in a real usage scenario, the transmitted package would be much larger, since according to the standard, additional bits are required for the transmission of the information, such as appropriate headings, redundancy bits for error correction etc. However, without loss of generality, for the needs of the present work, the selection of such a data package is totally adequate as it provides all the characteristics of a real signal.

4.3. Simulation Assumptions

In this work, we assume that the average illuminance that falls on a work desk is 500 Lux, since with this illuminance a person can work or study uninterruptedly [35]. Simulations were conducted considering three main characteristics: (a) the intensity of the incident light flux per unit area of the photodiode, (b) the information transmission frequency and (c) the power consumption of the entire AFE along with the pre-amplification stage. When one of the two characteristics is explored (light intensity or transmission frequency), constant conditions were chosen for the other one, by considering realistic values for the proper validation of the receiver.

Aiming to define the working range of the receiver under various operating conditions, the basic criterion for the first two factors is the signal integrity at the output of the AFE. The signal integrity is lost even when a single bit from the data sequence in Figure 8 is lost at the output of the AFE. By considering that the transition threshold at the input of the digital block that follows, which is driven by the AFE, is at $V_{dd}/2$, a bit is lost when an expected output signal transition does not pass this threshold.

4.4. Incident Illuminance Intensity on the Surface of the Photodiode

In order to determine the influence of the light power intensity on the receiver operation, the illuminance intensity value (Lux) was changed in the subcircuit of the photodiode by changing its photocurrent accordingly. A constant signal transmission frequency of 5 MHz was considered as a typical frequency of the system.

It was also observed that there is no differentiation between the three AFE variations when illuminance is increased, while low values of illuminance present an observable differentiation as described below.

For the first pre-amplification topology (I) (single BJT), as the incident illuminance in the photodiode decreases, the signal integrity is lost at 95 Lux, as shown in Figure 9. On the contrary, for the pre-amplification topologies (II) and (III), the signal integrity is lost at the illuminance of 5 Lux, as shown in Figures 10 and 11 respectively. Note that for the AFE topology without pre-amplification signal integrity issues arise below 120 Lux.

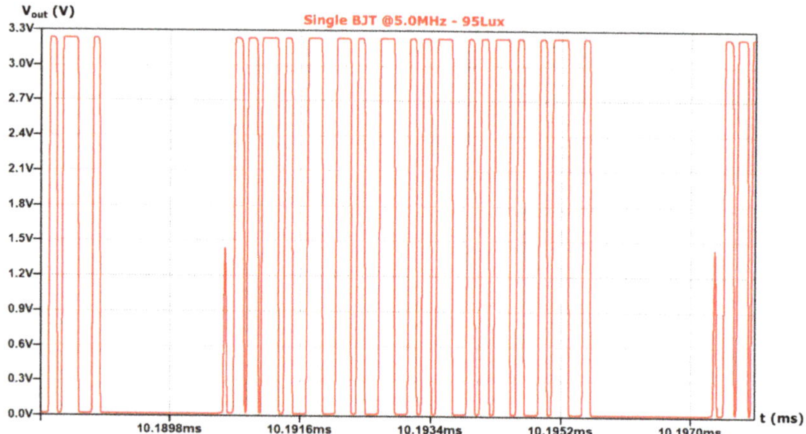

Figure 9. Output signal of the AFE with the pre-amplification topology (I) at 95 Lux.

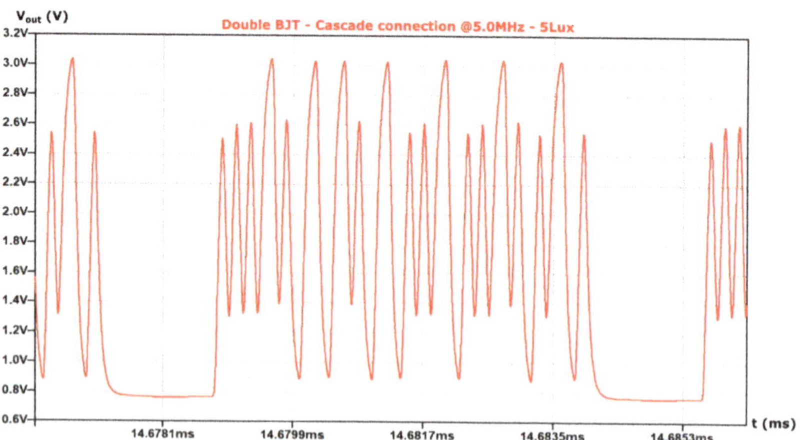

Figure 10. Output signal of the AFE with the pre-amplification topology (II) at 5 Lux.

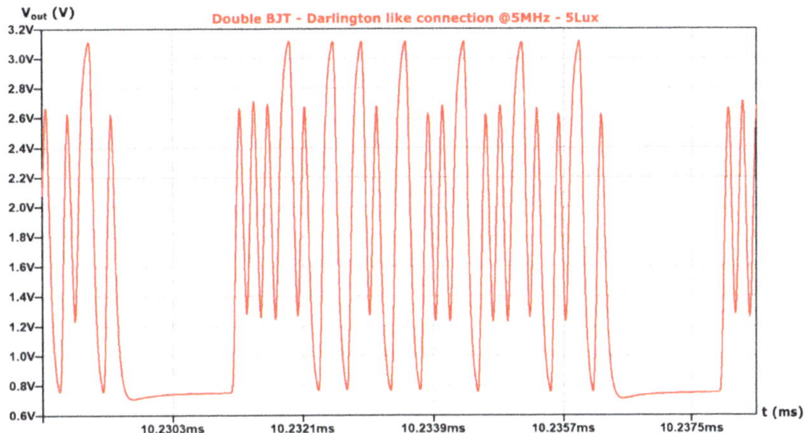

Figure 11. Output signal of the AFE with the pre-amplification topology (III) at 5 Lux.

Topologies (II) and (III) present similar behaviors and they are clearly superior over (I) when the sensitivity to the illuminance intensity is considered.

4.5. Data Transmission Frequency

Next, the data transmission frequency efficiency of each topology is considered. For this purpose, the light intensity (and consequently the corresponding photocurrent) was kept constant at the value of 500 Lux.

Unlike the previous simulation results, the proposed topologies presented no differentiation for lower frequencies, while they did when transmission frequency increased.

When the frequency varies, the signal remains intact for topology (I) up to the frequency of 7.9 MHz, as shown in Figure 12, while for topology (II) up to 11.25 MHz (Figure 13) and for topology (III) up to 12.05 MHz (Figure 14), respectively.

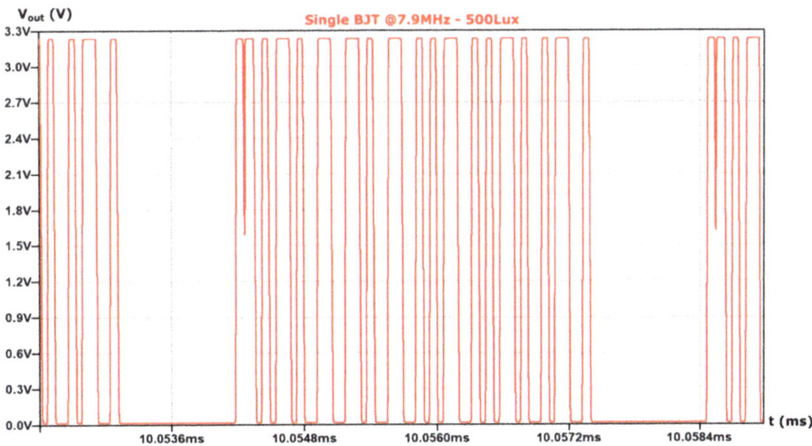

Figure 12. Output signal of the AFE with the pre-amplification topology (I) at 7.9 MHz.

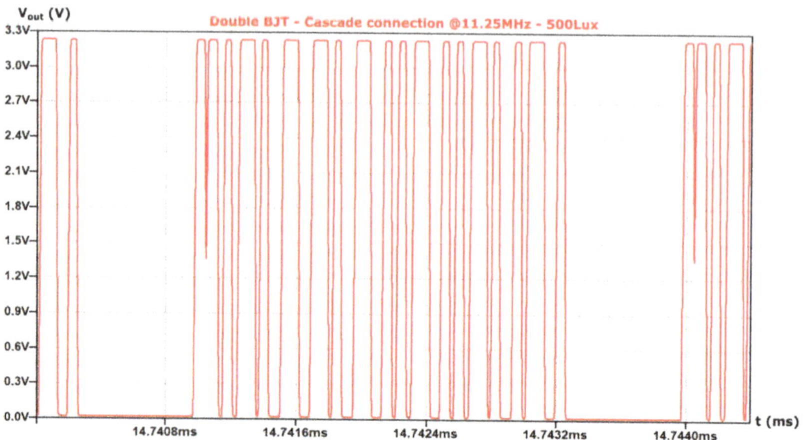

Figure 13. Output signal of the AFE with the pre-amplification topology (II) at 11.25 MHz.

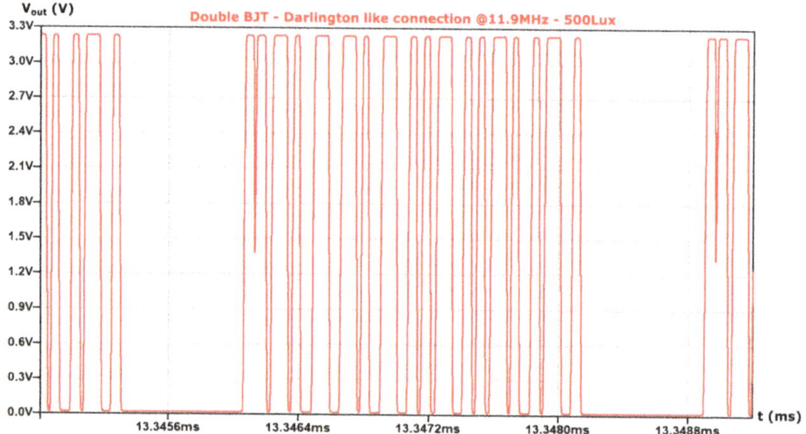

Figure 14. Output signal of the AFE with the pre-amplification topology (III) at 11.9 MHz.

Again, topology (I) seems to lag behind the other two topologies in terms of data transmission frequency. However, topologies (II) and (III) also present differentiations, with topology (III) being able to provide approximately 650 kHz higher data transmission frequency than topology (II).

4.6. Receiver Analog Front End (AFE) Power Consumption

Two types of simulations were performed on each AFE circuitry in order to examine the power consumption of the receiver in relation to the signal transmission frequency and the illuminance intensity. When the frequency is changed, the light intensity was set at 500 Lux, while when the light intensity is changed, the frequency was set at 500 kHz.

The average power consumption of the receiver AFE as a function of frequency for the three topologies (at 500 Lux illumination intensity) is presented in Figure 15.

As it can be observed in Figure 15, the frequency variation has little effect on the power consumption of each topology. More specifically, from 100 kHz to 5 MHz, the power consumption varies by 1.02% for topology (I), 1.003% for topology (II) and 1.006% for topology (III). However, with respect to topology (I), topology (II) presents a higher power consumption by 6.926% on average and topology (III) by 6.436% on average, respectively.

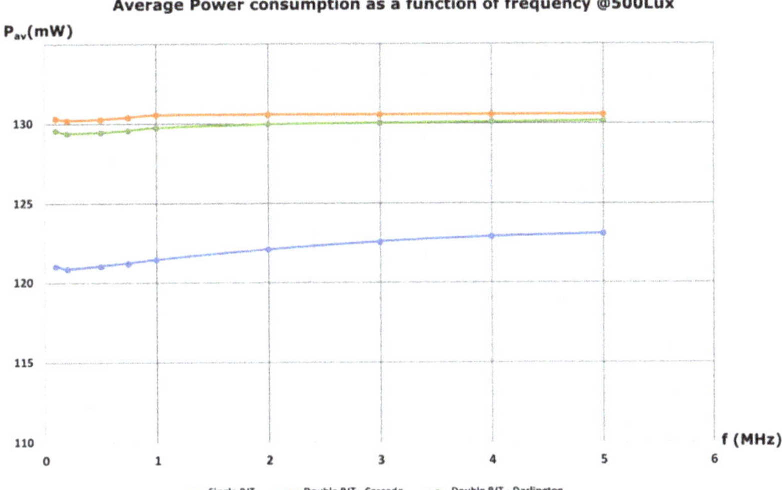

Figure 15. Average power consumption as a function of frequency for the receiver AFE with each one of the three pre-amplification topologies.

According to the above simulations on power consumption and considering the maximum operating frequency of each topology, the power consumption per MHz is extracted for the three topologies. Thus, the average power consumption per MHz reaches 16.18 mW/MHz, 11.78 mW/MHz and 11.13 mW/MHz, for topologies (I), (II) and (III), respectively (Figure 16).

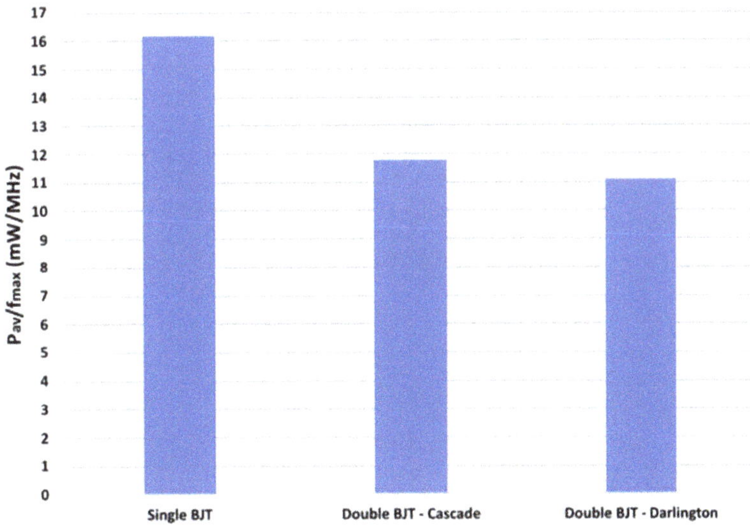

Figure 16. Power consumption per MHz for the receiver AFE with each one of the three pre-amplification topologies.

Moreover, the average power consumption of the receiver AFE as a function of illumination intensity for the three topologies (at a constant frequency of 500 kHz) is presented in Figure 17. The 500 kHz frequency was chosen aiming to reduce the simulation time, given that the frequency does not influence the power consumption, as discussed above.

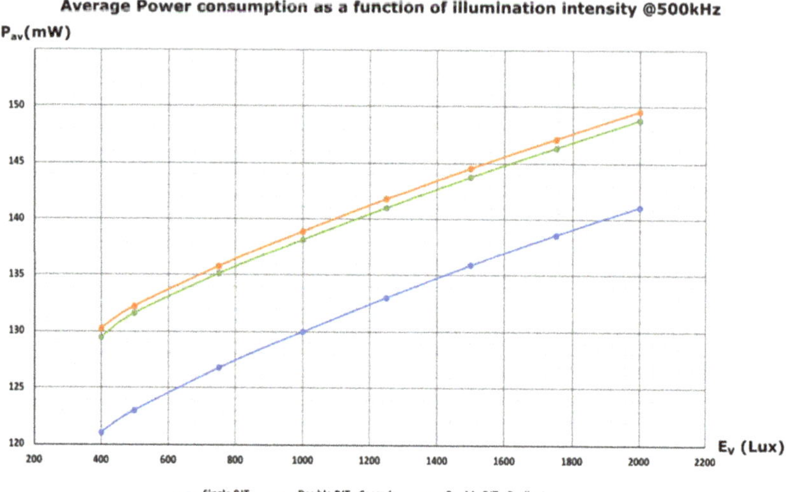

Figure 17. Average power consumption as a function of illumination intensity for the receiver AFE with each one of the three pre-amplification topologies.

In Figure 17 we observe that the variation of light intensity affects the power consumption. For each topology, it is obvious that the slope of the curves is similar. However, topologies (II) and (III) consume respectively on average 8.04% and 7.45% more power compared to topology (I).

5. Discussion and Conclusions

Although VLC is a promising technology there is a lack of extensive research activity on the receiver AFE design. Aiming to cover this area, we explore three pre-amplification topologies for the AFE of a VLC receiver, in order to improve the efficiency of this optical communication system, since typical AFE, without pre-amplification, are characterized by reduced signal integrity at low illuminance. In more detail, the typical AFE reaches an output voltage swing of 1.27 V at 5 MHz and 120 Lux. Referring to Figures 10 and 11, the AFE with either the pre-amplification topology II or III, has the same output voltage swing at the luminous flux of 5 Lux at the same transmission frequency. The light intensity of 5 Lux is 95.8% below the light intensity of 120 Lux and that means that the topologies II and III are more sensitive and capable to operate with smaller light signals than the typical AFE circuit without pre-amplification. Furthermore, the AFE with the pre-amplification topology I has a large, rail-to-rail, output voltage swing at 5 MHz and 95 Lux, which is 20.8% lower than the 120 Lux. All of the above indicate that the typical AFE is more prone to potential noise and less effective at low light signal intensities.

According to the simulations performed on the three topologies of the receiver AFE, it is stated that topology (III) provides the highest data transmission frequencies, 50.6% over topology (I) and 5.78% over topology (II). Topologies (II) and (III) support in parallel operations at low illuminance intensity, 94.74% below this of topology (I). On the other hand, topologies (II) and (III) present increased power consumption with respect to (I) by 7.5% and 6.97%, respectively. Moreover, topology (II) has almost the same response compared to topology (III) with respect to the sensitivity on the illuminance intensity. However, comparing these two topologies as a function of data transmission frequency, topology (III) performs better also providing a lower power consumption. By comparing the three topologies over the power consumption per MHz of operating frequency, once again topology (III) presents the best efficiency. More specifically, topology (II) has 5.87% more power consumption per MHz over topology (III), while topology (I) is characterized by 45.37% more power consumption per MHz compared to topology (III). Consequently, it is verified through the experimental results that topology (III) outperforms the other two topologies.

Author Contributions: Conceptualization, Y.S. and Y.T.; investigation, S.-I.P., G.P., C.P., Y.S., M.E.P., A.E., J.L. and Y.T.; methodology, S.-I.P.; project administration, A.E. and Y.T.; supervision, Y.T.; validation, S.-I.P., G.P., Y.S., A.E. and Y.T.; writing—review and editing, S.-I.P., Y.S., M.E.P., A.E., J.L. and Y.T. All authors have read and agreed to the published version of the manuscript.

Funding: This research has been co-financed by the European Union and Greek national funds through the Operational Program Competitiveness, Entrepreneurship and Innovation, under the call RESEARCH-CREATE-INNOVATE (project code: T1EDK-02419).

Institutional Review Board Statement: Not applicable.

Informed Consent Statement: Informed consent was obtained from all subjects involved in the study.

Data Availability Statement: Not applicable.

Conflicts of Interest: The authors declare no conflict of interest.

References

1. Khan, L.U. Visible light communication: Applications, architecture, standardization and research challenges. *Digit. Commun. Netw.* **2017**, *3*, 78–88. [CrossRef]
2. Tanaka, Y.; Komine, T.; Haruyama, S.; Nakagawa, M. Indoor Visible Light Data Transmission System Utilizing White LED Lights. *IEICE Trans. Commun.* **2003**, *86*, 2440–2454.
3. Haas, H. Wireless Data from Every Light Bulb; 1312294129; Technology, Entertainment, Design (TED) Conference: New York, NY, USA. Available online: https://www.ted.com/talks/harald_haas_wireless_data_from_every_light_bulb (accessed on 9 February 2022).
4. *802.15.7-2011*; IEEE Standard for Local and Metropolitan Area Networks—Part 15.7: Short-Range Wireless Optical Communication Using Visible Light. IEEE: Piscataway, NJ, USA, 2011. [CrossRef]
5. Wang, Q.; Giustiniano, D.; Gnawali, O. Low-Cost, Flexible and Open Platform for Visible Light Communication Networks. In Proceedings of the 2nd International Workshop on Hot Topics in Wireless, Paris, France, 11 September 2015; pp. 31–35.
6. Schmid, S.; Ziegler, J.; Corbellini, G.; Gross, T.R.; Mangold, S. Using consumer LED light bulbs for low-cost visible light communication systems. In Proceedings of the 1st ACM MobiCom Workshop on Visible Light Communication Systems, Maui, HI, USA, 7 September 2014; pp. 9–14. [CrossRef]
7. Chowdhury, M.Z.; Hossan, M.T.; Islam, A.; Jang, Y.M. A Comparative Survey of Optical Wireless Technologies: Architectures and Applications. *IEEE Access* **2018**, *6*, 9819–9840. [CrossRef]
8. Hansen, C. WiGiG: Multi-gigabit wireless communications in the 60 GHz band. *IEEE Wirel. Commun.* **2011**, *18*, 6–7. [CrossRef]
9. Tsonev, D.; Videv, S.; Haas, H. Towards a 100 Gb/s visible light wireless access network. *Opt. Express* **2015**, *23*, 1627. [CrossRef] [PubMed]
10. Komine, T.; Nakagawa, M. Fundamental analysis for visible-light communication system using LED lights. *IEEE Trans. Consum. Electron.* **2004**, *50*, 100–107. [CrossRef]
11. Azhar, A.H.; Tran, T.-A.; O'Brien, D. Demonstration of high-speed data transmission using MIMO-OFDM visible light communications. In Proceedings of the 2010 IEEE Globecom Workshops, Miami, FL, USA, 6–10 December 2010; pp. 1052–1056.
12. Azhar, A.H.; Tran, T.-A.; O'Brien, D. A Gigabit/s Indoor Wireless Transmission Using MIMO-OFDM Visible-Light Communications. *IEEE Photonics Technol. Lett.* **2013**, *25*, 171–174. [CrossRef]
13. Tian, Z.; Wright, K.; Zhou, X. Lighting up the Internet of Things with DarkVLC. In Proceedings of the 17th International Workshop on Mobile Computing Systems and Applications, St. Augustine, FL, USA, 23–24 February 2016; pp. 33–38.
14. Papadimitratos, P.; La Fortelle, A.; Evenssen, K.; Brignolo, R.; Cosenza, S. Vehicular communication systems: Enabling technologies, applications, and future outlook on intelligent transportation. *IEEE Commun. Mag.* **2009**, *47*, 84–95. [CrossRef]
15. Liu, C.B.; Sadeghi, B.; Knightly, E.W. Enabling vehicular visible light communication (V2LC) networks. In Proceedings of the Eighth ACM International Workshop on Vehicular Inter-Networking—VANET'11, Las Vegas, NV, USA, 19–23 September 2011; p. 41.
16. Li, L.; Hu, P.; Peng, C.; Shen, G.; Zhao, F. Epsilon: A Visible Light Based Positioning System. In Proceedings of the 11th USENIX Conference on Networked Systems Design and Implementation, Seattle, WA, USA, 2 April 2014; pp. 331–343.
17. Kuo, Y.S.; Pannuto, P.; Hsiao, K.J.; Dutta, P. Luxapose: Indoor positioning with mobile phones and visible light. In Proceedings of the 20th annual international conference on Mobile computing and networking, New York, NY, USA, 7 September 2014; pp. 447–458. [CrossRef]
18. *802.15.7-2018*; IEEE Standard for Local and Metropolitan Area Networks—Part 15.7: Short-Range Optical Wireless Communications. IEEE: Piscataway, NJ, USA, 2011. [CrossRef]

19. Ren, H.; Wang, Z.; Han, S.; Chen, J.; Yu, C.; Xu, C.; Yu, J. Performance Improvement of M-QAM OFDM-NOMA Visible Light Communication Systems. In Proceedings of the 2018 IEEE Global Communications Conference (GLOBECOM), Abu Dhabi, United Arab Emirates, 9–13 December 2018; pp. 1–6.
20. Retamal, J.R.D.; Oubei, H.M.; Janjua, B.; Chi, Y.-C.; Wang, H.-Y.; Tsai, C.-T.; Ng, T.K.; Hsieh, D.-H.; Kuo, H.-C.; Alouini, M.-S.; et al. 4-Gbit/s visible light communication link based on 16-QAM OFDM transmission over remote phosphor-film converted white light by using blue laser diode. *Opt. Express* **2015**, *23*, 33656. [CrossRef] [PubMed]
21. Chun, H.; Rajbhandari, S.; Faulkner, G.; Tsonev, D.; Xie, E.; McKendry, J.J.D.; Gu, E.; Dawson, M.D.; O'Brien, D.C.; Haas, H. LED Based Wavelength Division Multiplexed 10 Gb/s Visible Light Communications. *J. Light. Technol.* **2016**, *34*, 3047–3052. [CrossRef]
22. Kizilirmak, R.C.; Rowell, C.R.; Uysal, M. Non-orthogonal multiple access (NOMA) for indoor visible light communications. In Proceedings of the 2015 4th International Workshop on Optical Wireless Communications (IWOW), Istanbul, Turkey, 7–8 September 2015; pp. 98–101.
23. Sadat, H.; Abaza, M.; Gasser, S.M.; ElBadawy, H. Performance Analysis of Cooperative Non-Orthogonal Multiple Access in Visible Light Communication. *Appl. Sci.* **2019**, *9*, 4004. [CrossRef]
24. Simsek, C.; Tugcu, E.; Albayrak, C.; Yazgan, A.; Turk, K. Performance of Visible Light Communication with Colour Shift Keying Modulation and Polar Code. In Proceedings of the 2018 41st International Conference on Telecommunications and Signal Processing (TSP), Athens, Greece, 4–6 July 2018; pp. 1–4.
25. Dai, L.; Fang, Y.; Yang, Z.; Chen, P.; Li, Y. Protograph LDPC-Coded BICM-ID With Irregular CSK Mapping in Visible Light Communication Systems. *IEEE Trans. Veh. Technol.* **2021**, *70*, 11033–11038. [CrossRef]
26. Liao, C.-L.; Chang, Y.-F.; Ho, C.-L.; Wu, M.-C.; Hsieh, Y.-T.; Li, C.-Y.; Houng, M.-P.; Yang, C.-F. Light-emitting diodes for visible light communication. In Proceedings of the 2015 International Wireless Communications and Mobile Computing Conference (IWCMC), Dubrovnik, Croatia, 24–28 August 2015; pp. 665–667.
27. Mirvakili, A.; Koomson, V.J. High efficiency LED driver design for concurrent data transmission and PWM dimming control for indoor visible light communication. In *2012 IEEE Photonics Society Summer Topical Meeting Series*; IEEE: Seattle, WA, USA, 2012; pp. 132–133.
28. Gong, C.-S.A.; Lee, Y.-C.; Lai, J.-L.; Yu, C.-H.; Huang, L.R.; Yang, C.-Y. The High-efficiency LED Driver for Visible Light Communication Applications. *Sci. Rep.* **2016**, *6*, 30991. [CrossRef] [PubMed]
29. Fuada, S.; Putra, A.P.; Aska, Y.; Adiono, T. Trans-impedance amplifier (HA) design for Visible Light Communication (VLC) using commercially available OP-AMP. In Proceedings of the 2016 3rd International Conference on Information Technology, Computer, and Electrical Engineering (ICITACEE), Semarang, Indonesia, 19–20 October 2016; pp. 31–36.
30. Wielandt, S.; De Lausnay, S.; De Strycker, L. *Texas Instruments Innovation Challenge: Europe Analog Design Contest 2015 Project Report: ceLEDsTIal Positioning*; Texas Instruments: Dallas, TX, USA, 2015.
31. Radial Sidelooker, SFH 206 K | OSRAM Opto Semiconductors. Available online: https://www.osram.com/ecat/Radial%20Sidelooker%20SFH%20206%20K/com/en/class_pim_web_catalog_103489/prd_pim_device_2219558/ (accessed on 27 December 2021).
32. Available online: https://www.analog.com/media/en/technical-documentation/data-sheets/LTC6228-6229.pdf (accessed on 19 December 2021).
33. Available online: https://www.analog.com/media/en/technical-documentation/data-sheets/ada4622-1-4622-2-4622-4.pdf (accessed on 19 December 2021).
34. Low Profile Silicon NPN RF Bipolar Transistor. Available online: https://www.infineon.com/dgdl/Infineon-BFR340F-DS-v02_00-EN.pdf?fileId=5546d462689a790c01690f03e34d3934 (accessed on 19 December 2021).
35. British Standards Institution. *Light and Lighting—Lighting of Work Places. Part 1*; BSI: London, UK, 2011.

Article

Performance Analysis of 2D and 3D Bufferless NoCs Using Markov Chain Models

Konstantinos Tatas 

Department of Electrical and Computer Engineering and Informatics, Frederick University, Nicosia 1036, Cyprus; com.tk@frederick.ac.cy

Abstract: Performance analysis and design space exploration of bufferless Networks-on-Chip is done mainly through time-consuming cycle-accurate simulation, due to the chaotic nature of packet deflections, which have thus far prevented the development of an accurate analytical model. In order to raise the level of abstraction as well as capture the inherently probabilistic behavior of deflection routing, this paper presents a methodology for employing Markov chain models in the analysis of the behavior of bufferless Networks-on-Chip. A formal way of describing a bufferless NoC topology as a set of discrete-time Markov chains is presented. It is demonstrated that by combining this description with the network average distance, it is possible to obtain the expectation of the number of hops between any pair of nodes in the network as a function of the flit deflection probability. Comparisons between the proposed model and cycle-accurate simulation demonstrate the accuracy achieved by the model, with negligible computational cost. The useful range of the proposed model is quantified, demonstrating that it has an error of less than 10% for a significant proportion (between 33 and 75%) of the injection rate range below saturation. Finally, a simple equation for comparing mesh topologies with a "back-of-the-envelope" calculation is introduced.

Keywords: Network-on-Chip; bufferless routing; 3D NoC; Markov chains

Citation: Tatas, K. Performance Analysis of 2D and 3D Bufferless NoCs Using Markov Chain Models. *Technologies* **2022**, *10*, 27. https://doi.org/10.3390/technologies10010027

Academic Editor: Spiros Nikolaidis

Received: 31 December 2021
Accepted: 29 January 2022
Published: 2 February 2022

Publisher's Note: MDPI stays neutral with regard to jurisdictional claims in published maps and institutional affiliations.

1. Introduction

Networks-on-Chip have long been the dominant design paradigm for multi- and many-core architectures [1] and an active field of research for decades [2]. They emerged in the quest for a communication architecture that would provide the scalability required for high-performance heterogeneous systems [3]. Such systems are essential for the implementation of demanding applications such as cloud-based 3D real-time vision [4], convolutional neural networks [5], etc. Originally, NoC router architectures were quite similar to the ones found in off-chip interconnection networks [6]. Later efforts started taking into account the distinctive characteristics of the on-chip environment, in order to improve router performance, area and power consumption. One of the main such differences between on-chip and off-chip networks is the availability of links and buffers. NoC researchers focused on developing various router architectures unique to the on-chip environment in an effort to leverage the availability of links versus the scarcity of buffers. Many of these router architectures attempted to optimize the router buffer size and utilization [7–10].

On one extreme, Ref. [11] proposed eliminating buffers altogether from the router architecture, resulting in a bufferless router. Bufferless routers rely on deflecting packets that cannot be forwarded to a productive port (occupied by another packet) since packets cannot be stored. The result of this is that bufferless NoCs do not suffer from deadlock but may instead suffer from livelock, when the packets are continuously deflected from their path to the destination, adding to network congestion and causing more packet deflections. The packet deflections add hops to the path of the packet, which no longer follows the shortest distance to the destination. This leads to increased network latency, not because packets remain in buffers, unable to proceed towards their destination, but because they

distance from its destination). Secondly, the state that corresponds to the destination has probability one that it will transition to itself (absorbing state) since the flit has reached its destination. Thirdly, if a flit is deflected at the destination node router because another flit has priority for ejection, it must traverse a circular path to return to the destination router. Note that, in bufferless routing, it is impossible for a flit to remain at the same distance in two consecutive cycles due to the absence of buffering and, therefore, there is no transition from a state to itself except in the absorbing state.

Figure 2 illustrates the proposed methodology of encoding routing paths with deflections using the concrete example of Figure 1. The example shows the same transition probability in all states, and while clearly this is not perfectly true, as we demonstrate in our exploration of deflection probability in Section 4, simulations show small variation and our aim is to obtain the average latency across nodes; therefore, we can consider it as the average transition probability.

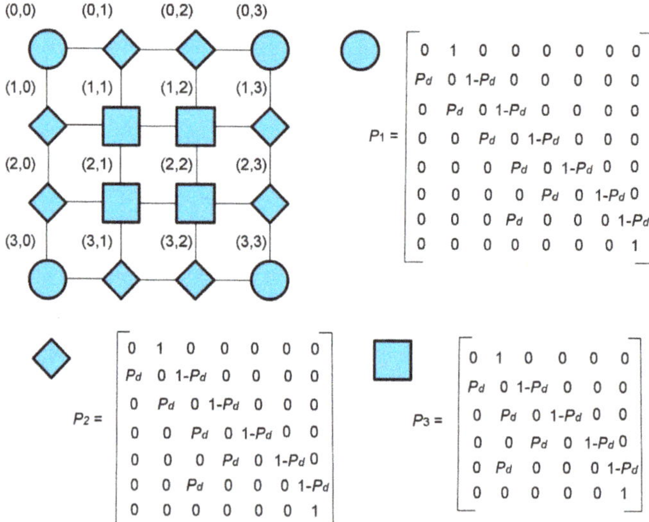

Figure 2. Markov chain transition matrices for a 4×4 2D NoC topology. The source−destination pair of Figure 1 corresponds to the third row of the second transition matrix. The three distance classes of nodes are indicated using different node symbols.

Definition 1. *The maximum shortest distance from a node (vertex) u in a network topology is the maximum among the minimum distances of all other nodes from u.*

This definition is consistent with [24] for a graph, but we consider individual nodes and do not extend it to the entire graph (network). It can also be seen as the minimum distance of the farthest node in the graph from u. Formally,

$$d_u = \max(d(u, v_i)), \text{ for } i = 1, 2, \ldots, N$$

For example, in the case of Figure 2, $d(0, 0) = d(0, 3) = d(3, 0) = d(3, 3) = 6$ (nodes indicated as circles). Each node has a unique maximum shortest distance, and depending on the regularity of the topology, multiple nodes may share the same maximum shortest distance. For a mesh topology, the possible maximum shortest distances depend on the diameter of the network. As shown in Figure 3, for a 4×4 mesh, there are three possible maximum distances for all sixteen nodes: four nodes have a maximum distance of six hops (network maximum), eight nodes have a maximum shortest distance of five hops, and the four inner nodes have a maximum shortest distance of four hopes. The maximum distance for each node can be used to separate the nodes in the topology into distance classes.

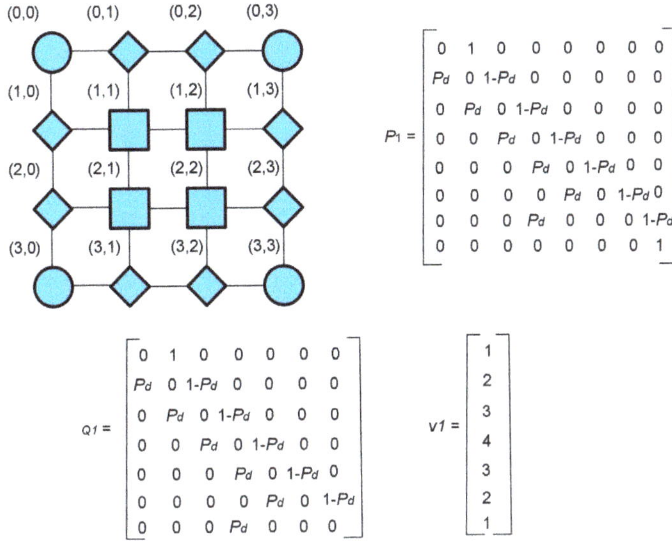

Figure 3. P and Q matrices for nodes marked as circles and corresponding node distances vector v_1.

Definition 2. *A distance class is a set of nodes in the topology that have the same maximum shortest distance.*

From the above definition, it is clear that the union of all distance classes is the set of all nodes in the network and that each distance class corresponds to a unique Markov transition matrix. Even though distance classes are technically sets and not proper classes, we prefer the term "class" in this context to emphasize the fact that they are defined by a common distance property.

After obtaining the Markov chain transition matrices for the given topology, known operations for the expectation can be used, which greatly simplify the analysis of the network latency. More specifically, the expectation in a Markov chain is given by adding all the elements in the corresponding row of matrix N given by [25]:

$$N = (I - Q)^{-1} \tag{1}$$

where Q is the transition matrix without the row and column of the absorbing state. N is called the fundamental matrix of the transition matrix P, and the expectation of the Markov chain is given by adding the elements of each row to obtain the expectation vector. In order to add the elements in each row, we multiply N with the vector j, and obtain the expectation vector:

$$ex = N_j \tag{2}$$

The expectation vector gives the expected latency for the class of destination nodes described by the transition matrix from all possible node distances in the network. However, there are likely different numbers of nodes that correspond to those particular distances based on the specific topology, as demonstrated in the example of Figure 2. For example, nodes $(0, 0)$, $(1, 1)$, $(2, 2)$ and $(3, 3)$ have two nodes from which they are one hop away, three nodes from which they are two hops away, four nodes from which they are three hops away, three nodes from which they are four hops away, two nodes from which they are five hops away and one node from which they are six hops away.

We define the node distances vector v of each matrix Q as the vector whose elements correspond to the number of nodes that have a particular distance (hops plus one) in the network from the nodes that correspond to Q (Figure 3). Thus, the 4×4 mesh topology (for a given deflection probability) has been completely defined by the set $T = \{P_1, P_2, P_3, v_1, v_2, v_3\}$. This is an additional advantage of the proposed methodology, wherein regular topologies are "compressed" in the proposed model. More formally, given a network $N = \{V, E\}$ where the set of vertices, V, corresponds to the nodes (PEs and routers), and E is the edges corresponding to the bidirectional links, we determine a minimum set of Markov Chains and node distance vectors so that $N \to T$.

An important advantage of the proposed methodology is that by modeling the NoC topology as a set of Markov Chains and distance vectors obtained by distance classes, any topology can be abstracted away and therefore the proposed methodology can easily describe both 2D and 3D networks and is not necessarily limited to mesh topologies.

3.2. Traffic Modeling

In order to obtain the expected latency, besides the network topology, the traffic pattern must also be modeled. The traffic pattern is essentially defined as the number of packets that each node sends to other nodes. It can be modeled as a probability distribution. For example, in uniform random traffic, each node sends to all other nodes with equal probability, and is therefore a uniform probability distribution. Essentially, we use the average distance model proposed in [21], formulated as vector and matrix operations in order to combine it with our model. Since this probability distribution determines the number of packets exchanged among source–destination pairs, we model it as a vector of weights w. The elements of the vector are essentially the probabilities of the distribution, normalized so that the inner product w·v divided by N is equal to one, in order to be consistent with the definition of probability. Then, the mean expected latency in the class of nodes corresponding to the particular transition matrix is given by the inner product of vector *ex* and vector v divided by the total number of source–destination pairs:

$$\overline{EX_i} = \frac{ex \bullet (w \circ v)}{N - 1} \tag{3}$$

where $\circ$ is the Hadamard (element-wise) product operator between two vectors, "$\bullet$" is the inner product operator and N is the total number of nodes in the network (sixteen for a 4×4 mesh).

Therefore, the mean expected latency for the entire network is given by the equation:

$$\overline{EX} = \frac{\sum_{i=1}^{m} W_i \overline{EX_i}}{N} \tag{4}$$

where N is the total number of nodes in the network, m is the number of distance classes in the topology, W_i is the number of nodes with a specific maximum distance, and EX_i is the expectation calculated for that class of nodes.

One of the most important insights of the proposed methodology is that, due to the symmetry of most common topologies, which are regular, it is not necessary to model every source–destination pair as a unique Markov chain. The steps of the methodology are as follows:

1. Given a NoC topology, determine node distance classes and therefore the minimum number of Markov Chains.
2. Fill each Markov chain transition matrix with the transition probabilities.
3. Use Equations (1)–(3) to obtain the expectations for each source–destination node pair.
4. Use Equation (4) to obtain the mean expected latency for the entire network based on the traffic pattern.

Determining the minimum number of Markov chains requires an analysis of the target topology. On one extreme, if nodes are arranged in a Spidergon [1], then they all have a minimum distance of one hop and a maximum distance of two hops. This can be modeled with a single Markov chain. On the other extreme, a highly irregular topology may require a Markov chain for each source–destination pair. To use the previous formulation, what must be determined is the cardinality of the set of all maximum shortest distances in the NoC. The cardinality of the set is equal to the minimum number of Markov chains required.

4. Experimental Results

Evaluating the accuracy of the proposed model requires comparison with the current state of the art. Since high-level models typically trade off accuracy for calculation or execution time, the model will be successful if it is less accurate than simulation but more accurate than the average distance model over as wide a range of injection rates as possible. Therefore, we perform cycle-accurate simulations and compare the latency observed in simulation with the one predicted by the proposed model and the average distance model, for a range of injection rates between near-zero and until saturation due to livelock begins. Using the equations in the procedure described in Section 3 requires a priori knowledge of the deflection probability P_d. Our use of cycle-accurate simulation is twofold: firstly to obtain insight into P_d, and secondly to compare the proposed model with simulation in order to evaluate the proposed model.

4.1. Deflection Probability Simulation

P_d is dependent on the flit injection rate γ and the traffic pattern. The first is the number of flits injected per clock cycle per node (flit distribution in time) and the second is the rate at which a source node sends to other nodes (flit distribution in space). As already mentioned, we take the traffic pattern into account by adopting the average distance model [21].

In order to tackle the deflection probability, we used simulation to gain insight into its relation with γ. We performed simulations using HNoCs [26], a cycle-accurate simulator. The simulator was modified to support bufferless routing and to measure the deflection probabilities in every router. This was done using the frequentist definition of probability. The deflection probability is measured by counting the deflected flits and total flits in every node and dividing the two. Using the mean deflection probability of the simulations in the proposed model would yield highly accurate results, but then it would defeat the point of the foregoing simulation. We generated heat maps, such as the ones in Figure 4, for a 4×4 NoC to measure P_d for every node and compare it with γ.

γ = 0.04   Pavg = 0.0474

xy	0	1	2	3
0	0.056105	0.036647	0.049090647	0
1	0.066153	0.051434	0.052597452	0
2	0.065285	0.046261	0.047334686	0
3	0.027213	0.029051	0.017738067	0

γ = 0.06   Pavg = 0.0584

xy	0	1	2	3
0	0.105861	0.070734	0.077033	0.103565
1	0.064004	0.058276	0.061314	0.070988
2	0.064385	0.050568	0.051136	0.050386
3	0.02806	0.030212	0.019225	0.01346

Figure 4. Deflection frequencies heat map for $4 \times 4 \times 1$ NoC with uniform random traffic and injection rates 0.04 and 0.06 flits/cycle/node. The average deflection probability is close to the injection rate for injection rates up to 0.12 flits/cycle/node for this topology.

We reached the conclusion, after repeated simulations using various topologies, that, for the low injection rates where the network is not saturated, the average P_d is approximately equal to γ, and we use γ as the deflection probability in our results in Section 3. Clearly, the deflection probability varies with the location of the node and the traffic pattern, as can be seen from Figure 4. In our simulations, we assume dimension order (xy) routing for flits when there are no deflections. However, directly using the deflection probability obtained from simulation would defeat the purpose of the model. This is currently a limitation of the proposed work, but, as will be shown in the next subsection, the assumption that the deflection probability is the same for all nodes and equal to the injection rate leads to accurate latency prediction until the network begins to saturate.

However, at injection rates where the network begins to saturate, the deflection probability depends on the previous state of the network (old flits that have not reached their destination). In this case, P_d rapidly grows higher than γ and we expect our model to start to diverge.

4.2. Average Latency Analysis

We evaluate the proposed model's predictive ability by comparing the expected latency given by the model with the same latency obtained by cycle-accurate simulation. We consider various topology, injection rate and traffic pattern combinations. We use the assumptions discussed previously that $P_d = \gamma$ and that P_d is the same for all states (average deflection probability).

Figures 5–7 show the average latency in hops for three typical 3D NoC topologies, $4 \times 4 \times 4$ (Figure 5), $8 \times 4 \times 2$ (Figure 6) and $8 \times 8 \times 1$ (Figure 7), for rising injection rates comparing simulation, the average distance model [21] and the proposed model under the assumption that $P_d = \gamma$, for uniform random traffic.

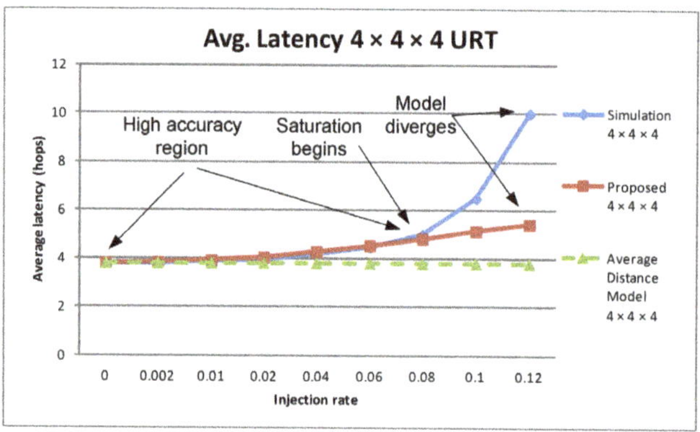

Figure 5. Average estimated latency vs. flit injection rate comparison between simulation, the proposed model and the average distance model for a $4 \times 4 \times 4$ NoC topology and uniform random traffic. The high-accuracy region, the beginning of saturation and the model divergence are clearly marked.

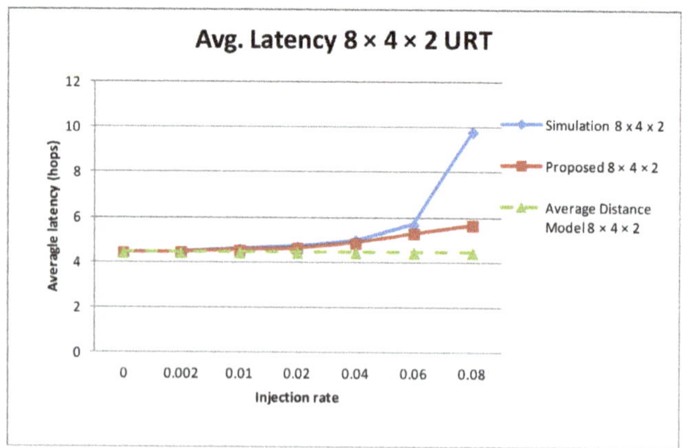

Figure 6. Average estimated latency vs. flit injection rate comparison between simulation, the proposed model and the average distance model for a $8 \times 4 \times 2$ NoC topology and uniform random traffic.

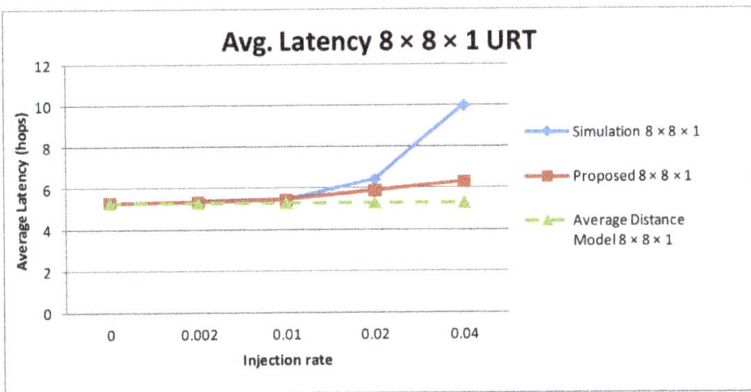

Figure 7. Average estimated latency vs. flit injection rate comparison between simulation, the proposed model and the average distance model for $8 \times 8 \times 1$ topology and uniform random traffic.

As shown in Figures 5–7, our results are almost surprisingly accurate in the network's high-performance injection rate region. Specifically, in the low injection rates, there is a high-accuracy region where the model estimate is very close to the simulation results. Above a certain injection rate threshold, γ_t, saturation begins. We refer to this value as the critical injection rate or as the saturation injection rate. The proposed model, even though it captures the non-linear relationship between the injection rate and the latency, nevertheless diverges for $\gamma > \gamma_t$ because the assumption of the deflection probability being approximately equal to the injection rate evidently no longer holds. However, the model does accurately predict the network latency over a wide range of injection rates before saturation and can be improved in the future by gaining more insight about the nature of deflections and adjusting the deflection probability. As expected, the proposed model's accuracy is higher than that of the average distance model and lower than that of the cycle-accurate simulation. Furthermore, the proposed model perfectly agrees with the average distance model for $P_d = 0$ (zero-load), and can therefore be viewed as a generalization of the average distance model.

Simulation results obtained using other synthetic traffic patterns are similar. This is not surprising since the proposed model indirectly takes traffic patterns into account in Equation (4), and, at low injection rates, the traffic pattern does not greatly affect the deflection probability as long as traffic remains non-bursty. Bursty traffic increases the average P_d and our assumption no longer holds. We intend to incorporate bursty traffic into our model in future work by exploring and quantifying the relation between the temporal distribution of traffic and Pd. In Figures 8–10, results are shown for a bit-complement traffic (BCT) pattern. More specifically, Figure 8 shows results for a $4 \times 4 \times 4$ topology and BCT, Figure 9 for an $8 \times 4 \times 2$ topology and BCT and Figure 10 for a $8 \times 8 \times 1$ topology and BCT.

The results for BCT highlight the same pattern: a high-accuracy region and then the gradual divergence between the hops predicted by the proposed model and simulation results. The BCT traffic patterns have a higher average distance, leading to saturation at lower γ_t, which is clearly dependent on both the topology and the traffic pattern. Furthermore, it seems that the more regular the topology, the higher the proposed model's accuracy. This is explored further in the next subsection.

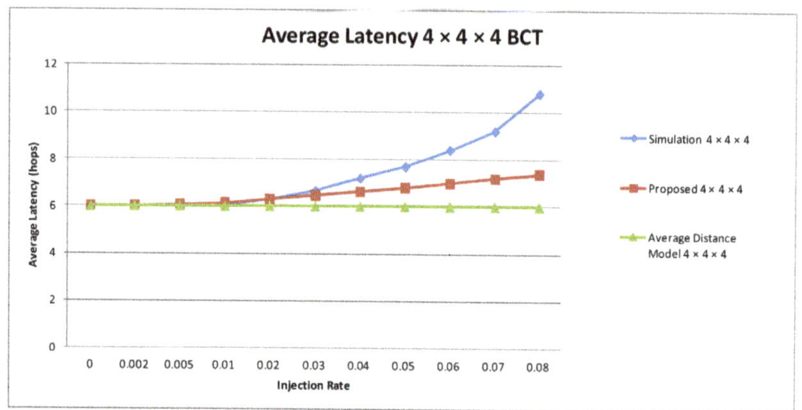

Figure 8. Average estimated latency vs. flit injection rate comparison between simulation, the proposed model and the average distance model for a $4 \times 4 \times 4$ NoC topology and bit-complement traffic.

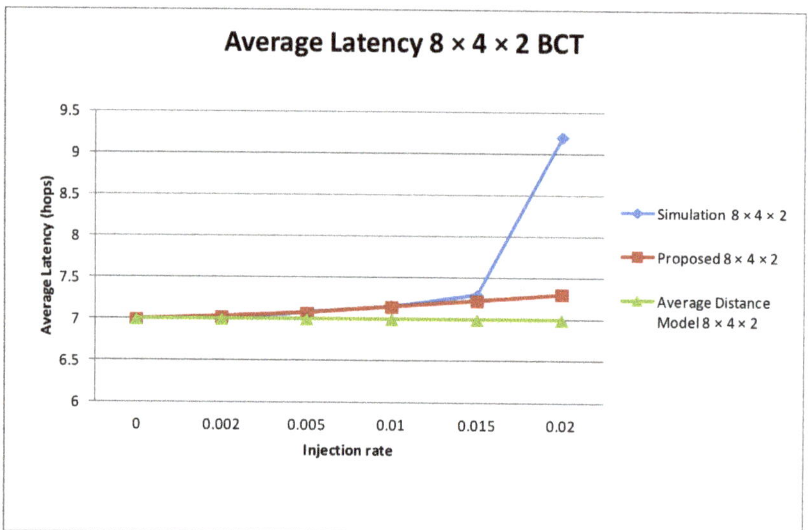

Figure 9. Average estimated latency vs. flit injection rate comparison between simulation, the proposed model and the average distance model for a $4 \times 4 \times 4$ NoC topology and bit-complement traffic.

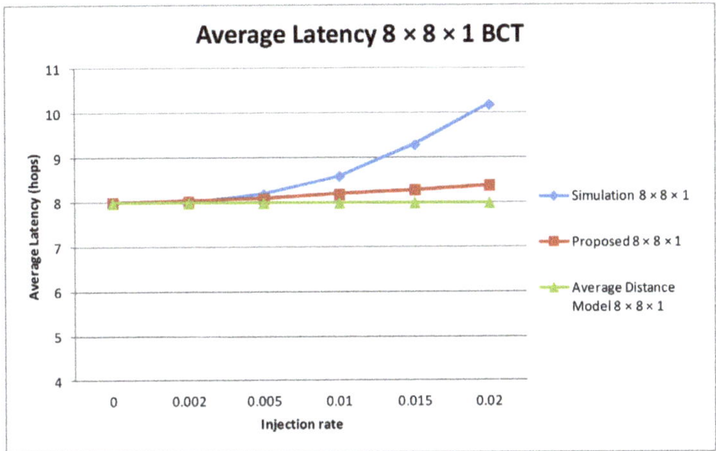

Figure 10. Average estimated latency vs. flit injection rate comparison between simulation, the proposed model and the average distance model for three NoC topologies and bit-complement traffic.

4.3. Model Accuracy Evaluation

Even though the above comparisons show promise, we attempt to further quantify and evaluate the usefulness of the proposed model. For this reason, we explore the accuracy as a function of the injection rate and compare it with the average distance model, with the simulation results assumed as perfectly accurate. We also attempt to quantify the injection rate range for which the model tends to be accurate, and therefore useful. Figure 11 plots the percentage absolute errors of the proposed model and the average distance model compared to simulation against the injection rate for URT and Figure 12 for BCT. The other topologies and traffic patterns follow similar trends and are summarized in Table 1.

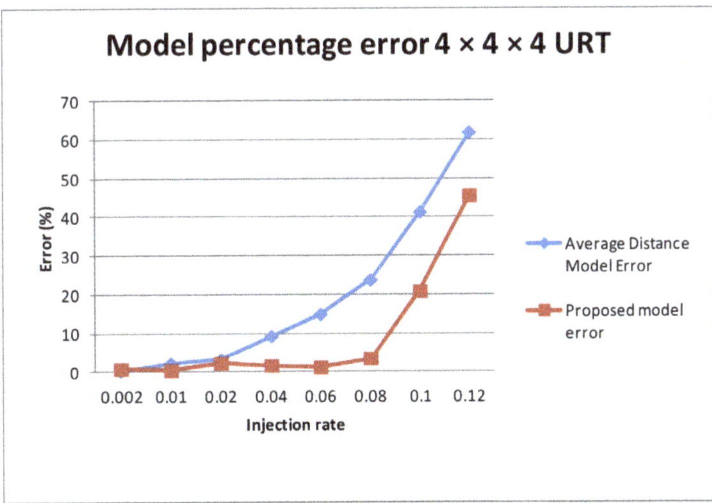

Figure 11. Percentage error vs. injection rate comparison between the proposed and the average distance model for 4 × 4 × 4 topology and uniform random traffic.

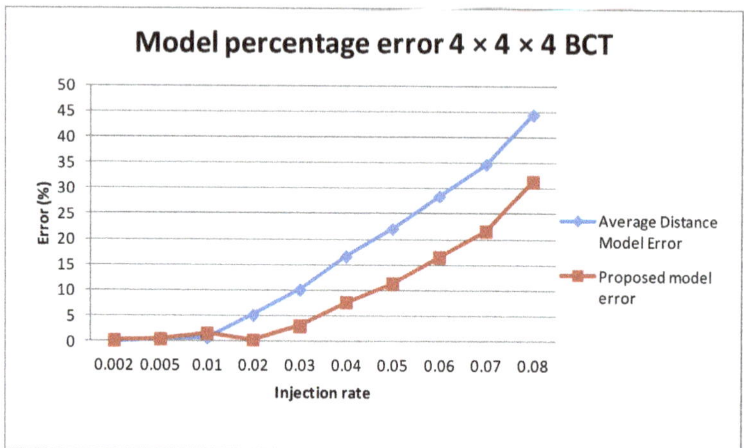

Figure 12. Percentage error vs. injection rate comparison between the proposed and the average distance model for $4 \times 4 \times 4$ topology and bit-complement traffic.

Table 1. Proposed model vs. simulation.

Topology	γ	Absolute Error URT/BCT	Percentage Error (%) URT/BCT	Normalized Error (%) URT/BCT
$4 \times 4 \times 4$	0.002	0.0945/0.0296	2.48/0.49	2.48/0.49
	0.01	0.0824/0.1005	2.11/1.66	2.16/1.67
	0.04	0.0434/0.5588	1.04/7.76	1.13/9.31
	0.06	0.1403/1.3748	3.34/16.60	3.68/23.24
	0.08	0.127	2.83	3.33
$8 \times 4 \times 2$	0.002	0.2385/0.0299	5.32/0.42	5.36/0.43
	0.01	0.2006/0.0695	4.36/0.92	4.51/0.99
	0.04	0.208	4.42	4.68
	0.06	0.14	2.8	3.15
	0.08	0.306	5.36	6.88
$8 \times 8 \times 1$	0.002	0.29/0.0379	5.47/0.47	5.54/0.47
	0.01	0.35/0.4073	6.46/4.74	6.61/5.09
	0.04	0.49	7.65	9.26

The above figures show that there is a slight oscillation in the proposed model error in the low injection rates, where both models exhibit low error. This is to be expected due to the randomness of high-level simulation. In the "medium" injection rates, the proposed model consistently exhibits lower error rates than the average distance model and yet both models exhibit a similar rising trend as the injection rate keeps increasing.

Table 1 summarizes in more detail the results illustrated in Figure 11 for $\gamma < \gamma_t$. Besides the percentage error, the absolute error and the normalized error are shown for URT. The absolute error is the absolute difference between the estimated and simulated latency, and the normalized error is the absolute error divided by the average distance in the topology.

The results of Table 1 illustrate the high accuracy of the model for $\gamma < \gamma_t$. Regarding the URT traffic pattern, for the $4 \times 4 \times 4$ topology, the normalized error is within 3.33% of simulation, for the $8 \times 4 \times 2$ topology within 6.88% and for the 8x8x1 within 9.26%. For the BCT traffic pattern, saturation begins at a lower γ_t and therefore the loss of accuracy begins more rapidly, and we do not show values above the saturation injection rate. Below the saturation rate, the accuracy is high in the case of BCT in most cases within 5%. The only exception is BCT in $4 \times 4 \times 4$, with a percentage error of 16.6 per cent. Even this error value is quite acceptable given the high-level nature of the model and the assumptions. Running on an Intel i7-7500U at 2.7 GHz using Octave, the estimation time is 30 to 60 milliseconds, several orders of magnitude faster compared to simulation, which takes from several minutes to even hours.

In order to explore the useful range of injection rates for the model, as well as gain insight into the relationship between topology, traffic pattern and saturation injection rate, we look for a metric that expresses how regular a topology is—in other words, how evenly spread are the nodes of the network in the mesh dimensions.

Thus, a simple metric for the topology regularity of a mesh with N nodes distributed in d dimensions is the ratio of the arithmetic over the geometric mean of the number of nodes in each dimension, defined as follows:

$$R = \frac{\sum_{i=1}^{d} n_i / d}{\left(\Pi_{i=1}^{d} n_i\right)^{1/d}} \tag{5}$$

where n_i is the number of vertices in dimension i, and d is the total number of dimensions.

By this definition, a topology with equal nodes in each dimension, such as $4 \times 4 \times 4$, has R equal to 1, since the arithmetic and geometric means of a set of equal numbers are equal. On the other hand, the greater the number of nodes in one particular dimension compared to the others, the higher the R metric, since the arithmetic mean is greater than or equal to the geometric mean. Specifically, $R(8 \times 4 \times 2) = (8 + 4 + 2)/3/4 = 7/6$ and $R(8 \times 8 \times 1) = (8 + 8 + 1)/3/4 = 17/12$. The reason that we do not use the more common inverted ratio of the geometric over the arithmetic mean instead is that, as will be shown, we intend to use R as a proxy for the average distance and, thus, it is easier to use in order to compare alternative topologies.

We define the injection rate for which a model error is less than 10% compared to simulation as the injection rate upper bound for this model γ_u and the useful range of the model as the ratio γ_u / γ_t. By this definition, we can quantify the usefulness of each model for a given topology and traffic pattern. We consider 10% to be a reasonable limit for accuracy for such a high-level model.

Table 2 explores the accuracy of the model compared to the injection rate in order to determine the useful range of the model, as well as to gain insight concerning the topology and traffic pattern relationship with the network saturation point as defined by the threshold injection rate γ_t. The topologies and traffic patterns are arranged in increasing average distance.

Table 2. Model error exploration.

Topology/Traffic Pattern	Average Distance $\bar{d}$	Topology Regularity (R)	γ_u ADM	γ_u Proposed	γ_t	Range ADM	Range Proposed
$4 \times 4 \times 4$/URT	3.8	1	0.04	0.09	0.12	33%	75%
$8 \times 4 \times 2$/URT	4.44	1.1667	0.04	0.06	0.08	50%	75%
$8 \times 8 \times 1$/URT	5.33	1.41667	0.015	0.02	0.06	25%	33%
$4 \times 4 \times 4$/BCT	6	1	0.03	0.05	0.08	37.5%	62.5%
$8 \times 4 \times 2$/BCT	7	1.1667	0.015	0.018	0.04	37.5%	45%
$8 \times 8 \times 1$/BCT	8	1.41667	0.009	0.011	0.025	36%	44%

Table 2 clearly shows that the proposed model extends the range of the average distance model by an average of 19.75% in the above topologies and traffic patterns. The lowest range extension is by 8% for $8 \times 4 \times 2$ and BCT traffic and the highest is by 42% for a $4 \times 4 \times 4$ topology and URT.

Another conclusion that can be drawn from Table 2 is the clear connection between the average network distance and the topology regularity as defined in Equation (5). In fact, closer inspection shows that the ratio of the average distance over the regularity for a given traffic pattern is almost identical for all specific topologies. In other words, multiplying the topology regularity by the average distance of the topology and traffic pattern with regularity equal to one yields the particular topology/traffic pattern average distance with high precision. Average distance and, therefore, topology regularity are also highly inversely correlated with γ_t. It can be seen that average distance alone cannot predict either γ_t or the useful range of the models. This relationship between average distance $\bar{d}$, topology regularity R and γ_t is further explored in Table 3.

Table 3. Relationship between average distance, regularity and saturation injection rate.

Topology/Traffic Pattern	$\bar{d}$	R	$\bar{d}$	γ_t
$4 \times 4 \times 4$/URT	3.8	1	3.8	0.12
$8 \times 4 \times 2$/URT	4.44	1.1667	3.8	0.08
$8 \times 8 \times 1$/URT	5.33	1.41667	3.76	0.06
$4 \times 4 \times 4$/BCT	6	1	6	0.08
$8 \times 4 \times 2$/BCT	7	1.1667	6	0.04
$8 \times 8 \times 1$/BCT	8	1.41667	5.65	0.025

As Table 3 illustrates, the topology regularity can be used to compare different mesh topologies in terms of zero-load latency and critical injection rate as effectively as the average distance. This is important because R is simpler to calculate, using Equation (3), than the average distance for a given mesh topology, even if it does not provide a sense of the actual number of hops. Essentially, the proposed work also provides a formula for simple "back-of-the-envelope" calculations for evaluating mesh NoC bufferless topologies. We stress, though, that this simple formula is limited to mesh topologies, while our methodology for modeling bufferless NoCs using Markov chains can be applied to any topology, regular or irregular.

Finally, the inverse correlation between R and γ_t is also clear, if harder to quantify. Preliminary calculations indicate that γ_t is approximately inversely proportional to the square of R for URT and the cube for BCT. However, for now, there is no approach other than simulation that can accurately determine the network saturation injection rate. This is also a limitation of the proposed model, which we plan to address in future work, as discussed in Section 5.

5. Conclusions and Future Work

A novel methodology for estimating latency in deflection routing networks based on stochastic processes is presented. The proposed methodology extends the zero-load average distance model by incorporating the probabilistic nature of deflection routing. The bufferless network is modeled as a set of Markov chains that capture the network topology. The proposed model is shown to be accurate for the range of injection rates where the network exhibits high performance, showing promising results as an approach. Furthermore, a simple equation reflecting the network topology regularity is proposed and its effectiveness in evaluating candidate topologies with simple "back-of-the-envelope" calculations is demonstrated.

In agreement with the maxim that "all models are wrong, but some models are useful", future work is aimed at tackling the existing limitations of the model. This leads to three directions: firstly, being able to predict the beginning of saturation where the model starts to diverge; secondly, gaining more insight into deflection probability, which would lead to more accurate estimation after saturation begins—this only requires a modification of the value of the transition probability in our model and can be readily applied; finally, we are currently exploring the effect of bursty traffic (the temporal distribution of traffic) on the deflection probability in order to incorporate it into our model.

Funding: This research received no external funding.

Institutional Review Board Statement: Not applicable.

Informed Consent Statement: Not applicable.

Data Availability Statement: Not applicable.

Conflicts of Interest: The author declares no conflict of interest.

Abbreviations

NoC	Network-on-Chip
URT	Uniform Random Traffic
BCT	Bit-Complement Traffic
ADM	Average Distance Model

References

1. Tatas, K.; Siozios, K.; Soudris, D.; Jantch, A. *Designing 2D and 3D Network-on-Chip Architectures*; Springer: New York, NY, USA, 2014.
2. Bohr, M.T. Interconnect scaling—The real limiter to high performance ULSI. In Proceedings of the International Electron Devices Meeting, Washington, DC, USA, 10–13 December 1995; pp. 241–244.
3. Swarbrick, I.; Gaitonde, D.; Ahmad, S.; Jayadev, B.; Cuppett, J.; Morshed, A.; Gaide, B.; Arbel, Y. Versal Network-on-Chip (NoC). In Proceedings of the 2019 IEEE Symposium on High-Performance Interconnects (HOTI), Santa Clara, CA, USA, 14–16 August 2019.
4. Ivanov, M.; Sergyienko, O.; Tyrsa, V.; Lindner, L.; Flores-Fuentes, W.; Rodríguez-Quiñonez, J.C.; Hernandez, W.; Mercorelli, P. Influence of data clouds fusion from 3D real-time vision system on robotic group dead reckoning in unknown terrain. *IEEE/CAA J. Automatica Sinica* **2020**, *7*, 368–385. [CrossRef]
5. Choi, W.; Duraisamy, K.; Kim, R.G.; Doppa, J.R.; Pande, P.P.; Marculescu, D.; Marculescu, R. On-Chip Communication Network for Efficient Training of Deep Convolutional Networks on Heterogeneous Manycore Systems. *IEEE TC* **2018**, *67*, 672–686. [CrossRef]
6. Guerrier, P.; Greiner, A. A Generic Architecture for On-Chip Packet-Switched Interconnections. In Proceedings of the Design, Automation and Test in Europe Conference and Exhibition (DATE), Paris, France, 27–30 March 2000.
7. Jafari, F.; Lu, Z.; Jantsch, A.; Yaghmaee, M.H. Buffer Optimization in Network-on-Chip Through Flow Regulation. *IEEE TCAD* **2010**, *29*, 1973–1986. [CrossRef]
8. Ramanujam, R.; Soteriou, V.; Lin, B.; Li-Shiuan, P. Design of a High-Throughput Distributed Shared-Buffer NoC Router. In Proceedings of the International Symposium on Networks-on-Chip (NOCS), Grenoble, France, 3–6 May 2010; pp. 69–78.
9. Wang, L.; Zhang, J.; Yang, X.; Wen, D. Router with Centralized Buffer for Network-on-Chip. In Proceedings of the Great Lakes Symposium on VLSI (GLSVLSI), Orange County, CA, USA, 6–8 June 2009; pp. 469–474.
10. Kodi, A.; Louri, A.; Wang, J. Design of energy-efficient channel buffers with router bypassing for network-on-chips (NoCs). In Proceedings of the 2009 10th International Symposium on Quality Electronic Design, San Jose, CA, USA, 16–18 March 2009; pp. 826–832.
11. Moscibroda, T.; Mutlu, O. A Case for Bufferless Routing in On-Chip Networks. In Proceedings of the 36th Annual International Symposium on Computer Architecture, New York, NY, USA, 11–15 June 2009; pp. 196–207.
12. Fallin, C.; Craik, C.; Mutlu, O. Chipper: A low-complexity bufferless deflection router. In Proceedings of the 17th IEEE International Symposium on High Performance Computer Architecture, San Antonio, TX, USA, 12–16 February 2011; pp. 144–155.
13. Feng, C.; Lu, Z.; Jantch, A.; Zhang, M. A 1-Cycle 1.25 GHz Bufferless Router for 3D Network-on-Chip. *IEICE Trans. Inf. Syst.* **2012**, *E95D*, 1519–1522. [CrossRef]
14. Feng, C.; Lu, Z.; Jantsch, A.; Zhang, M.; Xing, Z. Addressing transient and permanent faults in NoC with efficient fault-tolerant deflection router. *IEEE Trans. Large Scale Int. Syst. TVLSI* **2013**, *21*, 1053–1066. [CrossRef]
15. Tatas, K. High-performance 3D NoC bufferless router with approximate priority comparison. In Proceedings of the 7th International Conference on Modern Circuits and Systems Technologies (MOCAST), Thessaloniki, Greece, 7–9 May 2018.
16. Tatas, K.; Savva, S.; Kyriacou, C. 3DBUFFBLESS: A Novel Buffered-Bufferless Hybrid Router for 3D Networks-on-Chip. In Proceedings of the 27th International Symposium on Power and Timing Modeling, Optimization and Simulation (PATMOS 2017), Thessaloniki, Greece, 25–27 September 2017.
17. Audsley, N. Applying new scheduling theory to static priority pre-emptive scheduling. *Softw. Eng. J.* **1993**, *8*, 284–292. [CrossRef]
18. Qian, Z.L.Y.; Dou, W. Analysis of worst-case delay bounds for on-chip packet switching networks. *IEEE Trans. Comput. Aided Des. Integr. Circuits Syst.* **2010**, *29*, 802–815. [CrossRef]
19. Bekooij, M.; Hoes, R.; Moreira, O.; Poplavko, P.; Pastrnak, M.; Mesman, B.; Mol, J.D.; Stuijk, S.; Gheorghita, V.; van Meerbergen, J. Dataflow analysis for real-time embedded multiprocessor system design. In *Dynamic and Robust Streaming in and between Connected Consumer-Electronic Devices*; Springer: Berlin/Heidelberg, Germany, 2005; pp. 81–108.
20. Bogdan, P.; Marculescu, R. Non-stationary traffic analysis and its implications on multicore platform design. *IEEE Trans. Comput. Aided Des. Integr. Circuits Syst.* **2011**, *30*, 508–519. [CrossRef]
21. Weldezion, A.Y.; Grange, M.; Jantsch, A.; Tenhunen, H.; Pamunuwa, D. Zero-load predictive model for performance analysis in deflection routing NoCs. *Microprocess. Microsyst.* **2015**, *39*, 634–647. [CrossRef]
22. Tatas, K. Towards an Analytical Model of Latency in Deflection Routing: A Stochastic Process Approach for Bufferless NoCs. In Proceedings of the 10th International Conference on Modern Circuits and Systems Technologies (MOCAST), Thessaloniki, Greece, 5–7 July 2021.

23. Brémaud, P. *Markov Chains: Gibbs Fields, Monte Carlo Simulation and Queues*, 2nd ed.; Springer: Berlin/Heidelberg, Germany, 2020.
24. Selvi, T.; Vaidhyanathan. Maximum Distance in Graphs. *IJMTT* **2018**, *58*, 16–19. [CrossRef]
25. Haggstrom, O. *Finite Markov Chains and Algorithmic Applications*; Cambridge University Press: Cambridge, UK, 2002.
26. Ben-Itzhak, Y.; Zahavi, E.; Cidon, I.; Kolodny, A. HNOCS: Modular open-source simulator for Heterogeneous NoCs. In Proceedings of the International Conference on Embedded Computer Systems (SAMOS), Samos, Greece, 16–19 July 2012; pp. 51–57.

Reliable IoT-Based Monitoring and Control of Hydroponic Systems

Konstantinos Tatas [1], Ahmad Al-Zoubi [2], Nicholas Christofides [1], Chrysostomos Zannettis [3], Michael Chrysostomou [1], Stavros Panteli [4] and Anthony Antoniou [4,*]

1. Frederick Research Center, Frederick University, Nicosia 1036, Cyprus; com.tk@frederick.ac.cy (K.T.); eng.cn@frederick.ac.cy (N.C.); st009893@stud.frederick.ac.cy (M.C.)
2. Institute of Embedded Systems, Hamburg University of Technology (TUHH), Am Schwarzenberg-Campus 3 (E), 21073 Hamburg, Germany; ahmad.al.zoubi@tuhh.de
3. Department of Electrical and Computer Engineering and Informatics, Frederick University, Nicosia 1036, Cyprus; st016898@stud.frederick.ac.cy
4. Adaptive Hydroponics Limited, Larnaca 6010, Cyprus; adaptivehydroponics@gmail.com
* Correspondence: aant.gis@gmail.com

Abstract: This paper presents the design and implementation of iPONICS: an intelligent, low-cost IoT-based control and monitoring system for hydroponics greenhouses. The system is based on three types of sensor nodes. The main (master) node is responsible for controlling the pump, monitoring the quality of the water in the greenhouse and aggregating and transmitting the data from the slave nodes. Environment sensing slave nodes monitor the ambient conditions in the greenhouse and transmit the data to the main node. Security nodes monitor activity (movement in the area). The system monitors water quality and greenhouse temperature and humidity, ensuring that crops grow under optimal conditions according to hydroponics guidelines. Remote monitoring for the greenhouse keepers is facilitated by monitoring these parameters via connecting to a website. An innovative fuzzy inference engine determines the plant irrigation duration. The system is optimized for low power consumption in order to facilitate off-grid operation. Preliminary reliability analysis indicates that the system can tolerate various transient faults without requiring intervention.

Keywords: IoT; wireless sensor networks; hydroponics; smart agriculture

Citation: Tatas, K.; Al-Zoubi, A.; Christofides, N.; Zannettis, C.; Chrysostomou, M.; Panteli, S.; Antoniou, A. Reliable IoT-Based Monitoring and Control of Hydroponic Systems. *Technologies* **2022**, *10*, 26. https://doi.org/ 10.3390/technologies10010026

Academic Editors: Spiros Nikolaidis and Rodrigo Picos

Received: 29 December 2021
Accepted: 28 January 2022
Published: 2 February 2022

Publisher's Note: MDPI stays neutral with regard to jurisdictional claims in published maps and institutional affiliations.

1. Introduction

Water shortage and pollution are environmental threats affecting large parts of the EU, as well as the rest of the world. Climate change is expected to further increase the water shortage and threaten food production. This problem cannot be solved by a single effort but must be addressed on many levels. Food security is dependent on several environmental conditions, such as water quality and availability, soil condition and energy availability. Due to the extensive and exhausting practice of agriculture during the last decades, several environmental issues have arisen. Climate change is expected to affect water quality and quantity that are already in decline [1,2].

Smart farming is a capital-intensive and hi-tech system based on the application of modern information and communication technologies (ICTs) in agriculture. In IoT-based smart farming, a system is built for monitoring the crop field with the help of sensors and automating the irrigation system. Solutions for measuring environmental conditions and water quality based on electronic sensors and microcontrollers include ATLAS Scientific [3] and Libelium Waspmote [4]. Several research papers demonstrate such solutions, as shown in the surveys of [5,6]. Leveraging other important ICTs that have been successfully employed in other similar environments, such as big data [7], blockchain [8] and neural networks [9] is important for achieving precise and efficient agriculture [10].

Regarding IoT applications in agriculture, approaches range from open field cultivation [11,12] to greenhouses [13,14]. Typical parameters are temperature and humidity [11,12], but other conditions, such as light [13], are often monitored, as well as crop-specific parameters [11]. Greenhouse monitoring and control IoT systems are presented in [14,15]. In [13,16], the authors monitored parameters such as soil moisture, temperature, humidity and light in order to automate the irrigation system.

The above solutions are geared toward general greenhouse architectures or use generic parameters, such as temperature and humidity only. Hydroponic growth has additional monitoring requirements since the hydroponic irrigation system is different than the one used in soil cultivation. The complexity of hydroponic growth requires precise control, as well as monitoring parameters, such as pH and electrical conductivity, through specialized sensors and has only been automated in the case of large greenhouses with high cost, which is prohibitive to small-scale farmers. The iPONICS system attempts to leverage some of the technologies discussed above to provide a scaled-down, low-cost, yet innovative hydroponic solution that can be adopted by even hobby-type hydroponic systems. Specifically, we present the design and implementation of a novel, low-cost hydroponics monitoring and control system based on IoT technologies. In particular, the system is composed of a specialized wireless sensor network for monitoring the essential parameters for hydroponics and control for the irrigation process. It provides the user with a user-friendly web-based tool to monitor their crops, as well as being appraised by appropriate alarms and warnings. This greatly facilitates the observation of multiple hydroponics greenhouses with minimal effort and need for intervention. Unlike the systems discussed above, the proposed system targets hydroponic cultivation specifically.

Furthermore, we describe the design and implementation of a fuzzy inference engine (FIE) system for determining the system irrigation duration. A fuzzy set is defined by a membership function that can be any real number in the interval [0, 1], expressing the grade of membership for which an element belongs to that fuzzy set. The concept of fuzzy sets enables the use of fuzzy inference, which, in turn, uses the knowledge of an expert in a field of application to construct a set of "IF–THEN" rules. Fuzzy logic becomes especially useful in capturing a human expert's or operator's qualitative control experience into the control algorithm using linguistic rules. Fuzzy logic control (FLC) has been applied successfully for controlling numerous systems in which analytical models are not easily obtainable or the model itself, if available, is too complex and possibly highly non-linear in several applications ranging from network routing [17] to task mapping and scheduling [18,19].

The rest of the paper is organized as follows: Section 2 describes the background and Section 3 describes the proposed system in detail. The paper concludes with Section 4, which summarizes the results and discusses future work.

2. Background and System Requirements

Hydroponics [20] relies on fertilized and aerated water, which provides both nutrition and oxygen to a plant's root zone. It often involves sophisticated mechanization processes, which can be daunting to casual hobbyists, as well as small-scale commercial farms. Nutrient solutions must usually be below the temperature at which pathogen growth can begin, yet not so cool that root activity is suppressed. In hydroponics, as in conventional agriculture, nutrients should be adjusted to satisfy Liebig's law of the minimum for each specific plant variety [21].

As it can be seen from Table 1, certain crops are tolerant to a wide range of conductivity and pH values, such as spinach, asparagus and tomato, while others, such as strawberry, are very sensitive, especially to pH changes. This requires close monitoring of the above variables to adjust the nutrients.

Especially in low-cost hydroponics systems, the monitoring is done manually using specialized pH/EC meters, which is time consuming and requires constant proximity to the greenhouse. The iPONICS system facilitates remote monitoring of the values at low cost. Since it takes time for the plants to absorb the nutrients, a low sampling rate (as low as once per day) is sufficient for the water quality parameters, such as electrical conductivity and pH. In practice, we use a higher sampling rate to compensate for sensor and communication errors, as discussed in Section 4. Ambient temperature and humidity in the greenhouse are also monitored to ensure optimal growth.

Table 1. Crops' ideal pH and electrical conductivity values [21].

Crops	EC (mS/cm)	pH
Tomato	2.0–4.0	6.0–6.5
Cucumber	1.7–2.0	5.0–5.5
Strawberry	1.8–2.2	6.0
Banana	1.8–2.2	5.5–6.5
Spinach	1.8–2.3	6.0–7.0
Asparagus	1.4–1.8	6.0–6.8

Furthermore, since the purpose of the system is to achieve remote monitoring, additional monitoring requirements besides water quality arise. Specifically, the ability to generate alarms in cases of unexpected and possibly catastrophic conditions must be considered. To that purpose, we need frequent monitoring of ambient temperature to generate an alarm in the case of a fire in the greenhouse. For that reason, the ambient temperature is read every 8 seconds, but as long as it is normal, it is only recorded hourly. Finally, since the greenhouse containing the equipment can be situated at a remote location, we have need of some rudimentary security. While security is not essential to crop growth per se, we need to be at least notified of, if not prevent, unauthorized entry. The most important iPONICS system requirements are summarized in Table 2.

Table 2. iPONICS system requirements.

Requirement	Sampling Period	Alarm/Warning	Action
Monitor water quality (pH, temperature, EC, DO)	Adaptive, at least daily	Warning depending on crops	Automatically adapt irrigation duration
Monitor ambient temperature	8 s, record hourly	Alert (possible fire), SMS	System halt
Monitor ambient humidity	Hourly	Warning depending on crops	User intervention
Detect unauthorized entry	Event driven (interrupt)	Alert on server, SMS	User intervention
Put nodes to sleep to conserve energy	N/A	N/A	N/A

3. iPONICS System Design

Given the requirement of monitoring the temperature and humidity inside the greenhouse and the great variation in area between various types of greenhouses, the environment sensing system must be distributed across various nodes, the number of which will depend on the specific greenhouse dimensions. On the other hand, for monitoring the water quality and controlling the pump, one central node is sufficient. The iPONICS system concept is shown in Figure 1.

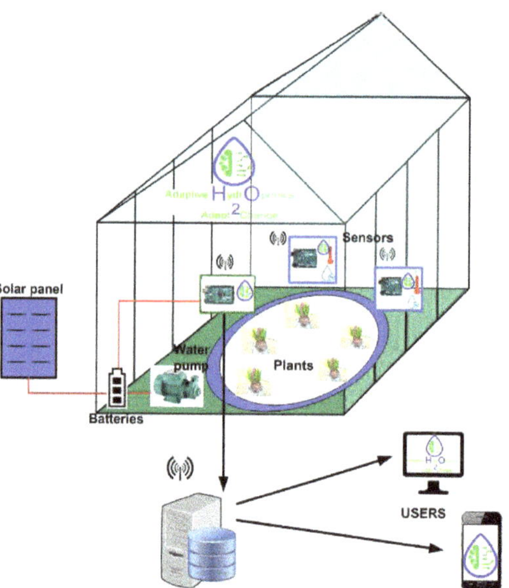

Figure 1. iPONICS system concept.

The iPONICS sensing and control system is a wireless sensor network with a star topology, as shown in Figure 2.

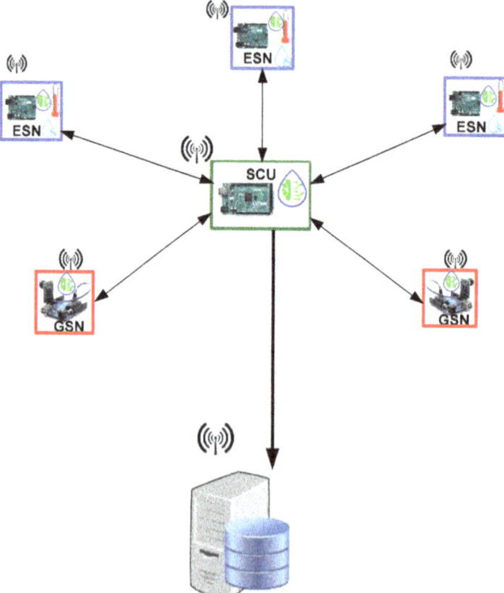

Figure 2. iPONICS WSN topology.

The system is composed of three types of nodes: the sensing and control unit (SCU), the environment sensing unit (ESN) and the greenhouse security unit (GSN). The greenhouse unit (GHU) is a greenhouse containing one SCU, which is the master node and several ESN and GSN "slave nodes", the number of which depends on the dimensions of the greenhouse.

3.1. Sense and Control Unit (SCU)

The SCU is the main (master) node, as well as the central node in the star topology of the WSN. It is responsible for measuring the circulating water quality through four sensors, namely, temperature, pH, water electrical conductivity (EC) and dissolved oxygen (DO). Furthermore, it is responsible for controlling the pump in an efficient manner by conserving energy while maintaining the required water flow for crops growth and requesting and receiving data from the slave nodes.

The sensors used were from Atlas Scientific [22–25]. The reasons for selecting these particular sensors were the following: First, they require infrequent recalibration, which is convenient while being deployed in a greenhouse during plant growth, which requires weeks to months. Second, they are easily integrated into an Arduino-based microcontroller system, reducing the development time and time-to-market. Third, they are long lasting at a reasonable cost, which are essential attributes in order to keep the overall cost of the system an attractive investment for farmers.

The SCU processing unit used was an Arduino Mega 2560, providing the required processing power, program memory and number of pins to support the following connections: the Xbee shield, the RTC shield, the Atlas Scientific EZO circuits, the GSM shield, the voltage sensor and, finally, the relay module to control the ON/OFF switching of the water pumps. All the shields and modules were supplied by the power drawn out of the Arduino by the 5v and GND pins. Figure 3 depicts the schematic of the SCU, while Figure 4 illustrates its operation using a flowchart.

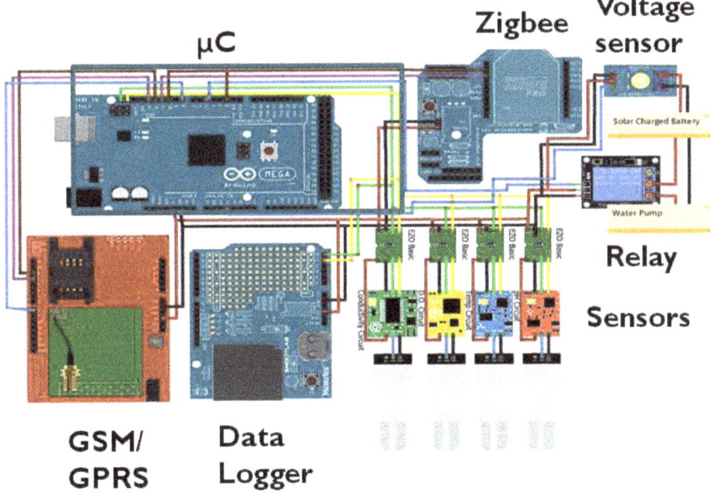

Figure 3. Sense and control unit annotated schematic diagram.

As shown in Figure 4, the SCU starts by initializing the serial and I2C communications and verifying that all modules are connected. In order to preserve energy, the microcontroller is put to sleep, along with the GSM shield since it is the highest power-consuming module. The microcontroller awakes in the following cases:

1. An "Alarm_Interrupt" that signals abnormal and possibly hazardous temperature/humidity readings in which the SCU activates the GSM to send an alarm SMS/email and upload these readings to the server. In this case, the system demands operator intervention and does not return to sleep or activate the pump.
2. A "Security_Interrupt" that signals unauthorized access to the site in which the SCU activates the GSM to send a security SMS/email and upload the captured images to the server before returning to sleep.

3. The SCU reaches the sampling time of water quality sensors in which it collects those readings temporally in the EEPROM. Afterwards, it activates the GSM shield and proceeds uploading the water quality readings, and then wakes up one ESN at a time and uploads its readings to the server. It returns to sleep afterward.
4. The SCU wakes up at fixed intervals for the pump control in which the state of the battery, along with the latest water quality readings, dictate the powering decision before returning to sleep.

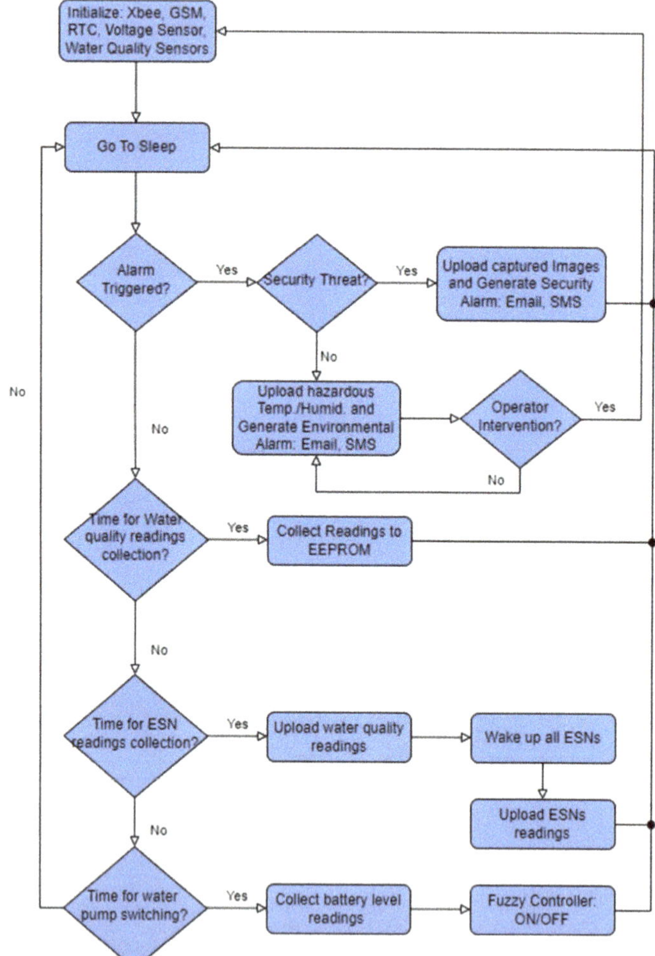

Figure 4. SCU flowchart.

Since the entire system is meant to be off-grid, it is crucial to minimize the power consumption of the units connected to the solar-charged battery whenever possible to guarantee uptime. The use of the sleep mode and intelligent control of the water pumps ensures moderate usage of the power resources. We measured the current drawn by the SCU under the operating conditions listed in Table 3 using both a multimeter and a current clamp. As it can be seen in Table 3, the SCU consumed around 1.25 W while sleeping, where the GSM was off and the microcontroller was in deep sleep mode. Other connected modules were still powered including the Xbee shield, as interrupts are asynchronous. In the idle state, the Arduino was awake to regularly do the water quality readings or control the water pump, where it consumed about 1.35 W. The power consumption almost doubled when the GSM shield was activated for transmission, where it consumed about 2.75 W. This could rise to 11.5 W given that the transmission is occurring under extreme weather conditions or a weak network connection, which caused the GSM to draw higher currents. On the positive side, without interruption, the GSM stayed off until the upload time, which happened a few times a day. Figure 5 shows an SCU prototype.

Table 3. SCU power ratings.

State	Current (A)	Power (W)
Sleep_mode	0.250	1.25
Idle	0.270	1.35
Transmission_idle	0.550	2.75
Transmission_extreme	2.300	11.5

Figure 5. SCU early prototype.

The maximum power (last entry in Table 3) was only consumed when the GSM module is first registered to the network. During normal operation, the SCU slept and consumed 1.25 W while consuming 2.75 W when transmitting data. Therefore, the SCU energy consumption per cycle was

$$E = 1.25T + 2.75t_e \quad \text{(mJ)} \tag{1}$$

where T is the sleep period and t is the execution time of the main loop (Appendix A). Essentially the sleep mode power was multiplied by the duration of the sleep period and the transmission power was multiplied by the time it took to transmit the data.

The execution time was measured in milliseconds by counting the processor cycles. The execution of the main loop with no ESNs present was 19,170 ms. With one ESN, the main loop execution time was 145,090 ms. Since all ESNs transmitted the same amount of data, the execution time was given by the formula:

$$t = 19{,}170 + 125{,}920 \times \text{ESN} \quad \text{(ms)} \tag{2}$$

Equation (2) also indirectly defines an upper bound in the sampling rate of the system, depending on the number of ESNs available. However, such a high sampling rate is not required since it takes at least hours for the water quality parameters to significantly change (when plants absorb nutrients) and, at most, days when most nutrients have been absorbed.

3.2. Environment Sensing Node (ESN)

Environment sensing nodes are slave nodes that are responsible for monitoring greenhouse environmental conditions, such as the ambient temperature and humidity. The necessity behind the ESNs is that abnormal temperature and humidity could imply a hazard, such as a fire. Therefore, the temperature is monitored every few minutes, but temperature and humidity readings are permanently stored on an hourly basis in an SD card. The ESN is interrupted by the SCU and then transmits the data to the SCU using Zigbee unless there is an alarm (high temperature), in which case, the ESN does not wait for an interrupt. This allows for effective monitoring of the greenhouse environment while minimizing data transmission and, therefore, power consumption. The number of ESNs present in each greenhouse is dependent on the greenhouse dimensions. Each ESN has its own ID number and Zigbee address. Figure 6 depicts the schematic diagram of the ESN circuit.

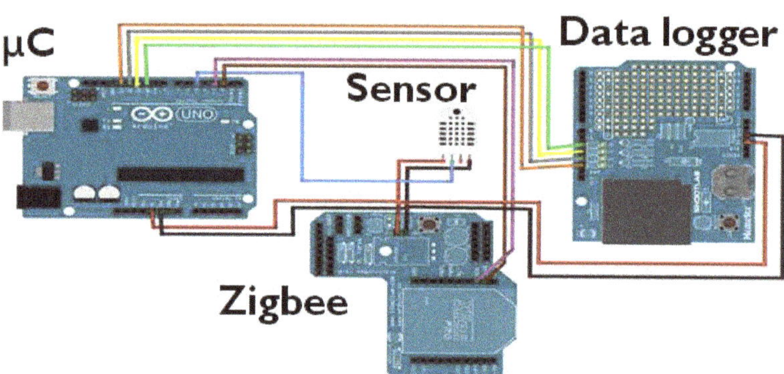

Figure 6. Environment sensing node annotated schematic diagram.

The ESN initialization procedure includes initializing the serial and SPI communication (Figure 7) and verifying the connection to all modules. After initialization, the ESN awakes on fixed intervals to collect the ambient temperature and humidity readings and record them on the SD card. If no abnormal readings are present, the ESN remains asleep until interrupted by the SCU when it is time for data collection. On the occurrence of the abnormal readings, the ESN interrupts the SCU with a Zigbee packet payload "Alarm_Interrupt_XX". The "XX" is a code representing either "FIRE" or "Connection" to distinguish whether the alarm is due to high temperature due to a possible fire in the greenhouse or just a connection problem.

The ESN power consumption in sleep mode was approximately 0.4 W compared with approximately 0.55 W during data transmission. However, as mentioned earlier in the SCU subsection (Section 3.1), the transmission only occurs a few times a day, unless there is an alarm. In both modes, the Xbee module stays active as it is an end device to be interrupted by the SCU.

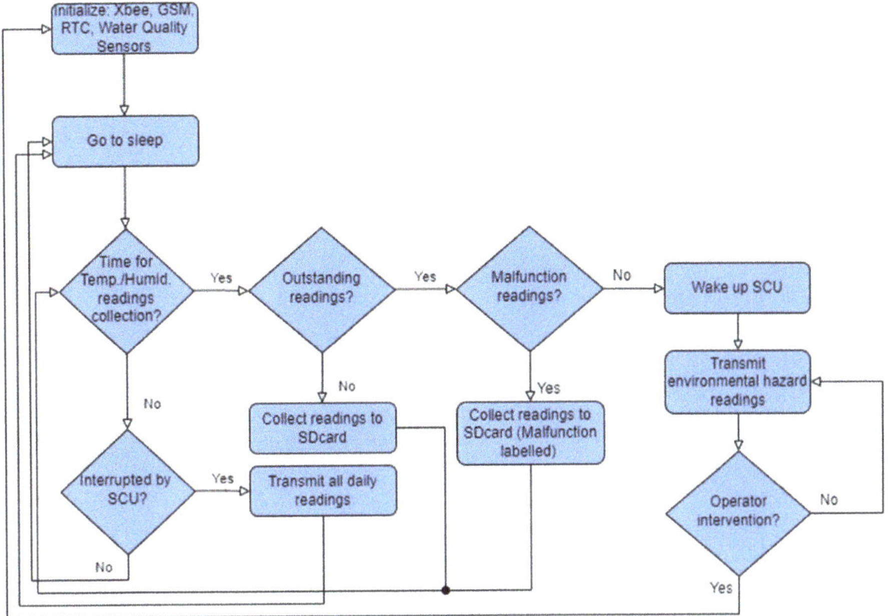

Figure 7. ESN flowchart.

3.3. Greenhouse Security Node (GSN)

As discussed in Section 2, the GSN was developed as a rudimentary security node and is not essential for the operation of the iPONICS system. It comprises a microcontroller with a motion sensor, camera, SD card and Zigbee transmitter/receiver. The reason such a node may be used is that a greenhouse could be in an isolated area, and we wish to prevent unauthorized entry. It is not meant to be a robust security solution, merely a low-cost warning system. The node spends the majority of time sleeping to conserve energy under normal circumstances. It is woken up by an external interrupt from the motion sensor, in which case, it activates the camera to take pictures and sends an alarm to the SCU through Zigbee. In the case of an authorized user accessing the system, the RFID is used to cancel the alarm. GSNs can also be deactivated and reactivated when receiving a command from the SCU. The microcontroller used for the implementation of the GSN was Arduino UNO, similar to the ESN. Figure 8 depicts the schematic diagram of the node's circuitry.

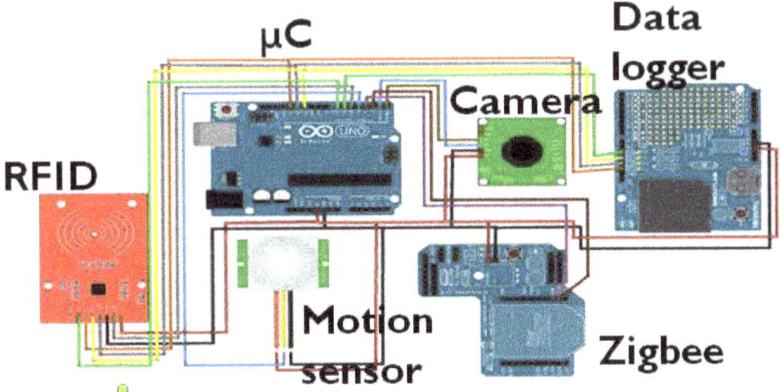

Figure 8. Greenhouse security node annotated schematic diagram.

Figure 9 shows a flowchart of the GSN operation. As shown in Figure 9, the node starts by initializing the I2C and SPI communications and ensures proper wiring of the modules. Then, the node falls into sleep mode and remains in that state until motion is detected. After such an event, the node expects an authorized ID tag swap. If no such verification of the identity of the user occurs in the next few seconds, the GSN interrupts the SCU and signals a "Security_Interrupt" inside the first payload, followed by the images that are data captured by the node. The GSN draws power of about 0.53 W in sleep mode and about 0.68 W during transmission depending on the distance from the SCU.

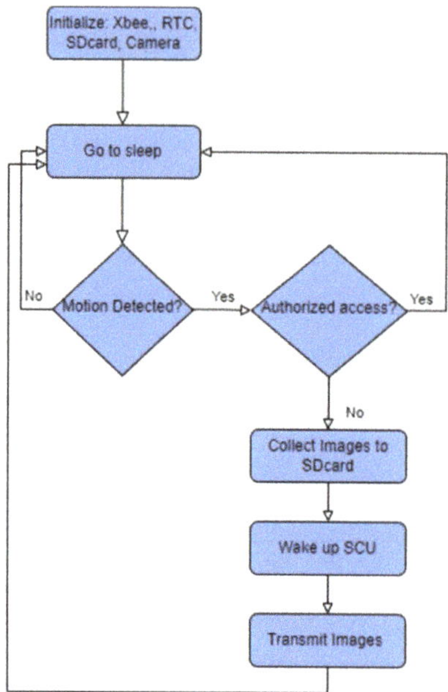

Figure 9. GSN flowchart.

3.4. SCU Fuzzy Inference Engine

An important novelty of the iPONICS method is the SCU FIE [26]. We describe here the design of the control algorithm separately from the rest of the SCU for clarity. There are several reasons for using FLC in iPONICS. First, we have to manage imprecision from sensor errors. Second, the ideal values for pH and electrical conductivity vary slightly between plants (Table 1), and we can accommodate several crops with the same controller by fuzzifying the inputs.

The architecture of the FIE is shown in Figure 10. It accepts three inputs: pH, EC and power supply voltage level. Since the power supply is from the battery being charged by a solar cell, the voltage level reflects the battery level according to the voltage/charge curve. The DO and water temperature are not used for two reasons: first, they are not documented in the literature to have a significant effect on efficient hydroponic culture, and second, using five or six outputs would require a large number of rules, most of which would be redundant. The FIE determines the irrigation duration based on the critical hydroponic parameters and the available energy. The general premise behind the knowledge base of the proposed scheme is to provide the available nutrients to the plants while being conservative with watering when energy and nutrients are low.

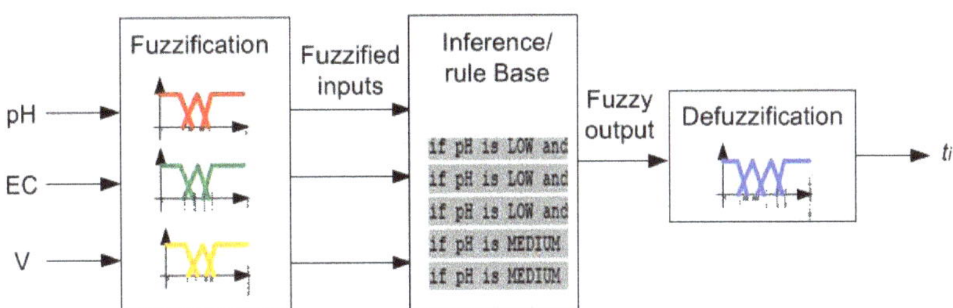

Figure 10. Fuzzy inference engine architecture.

According to hydroponics guidelines, the irrigation duration is given as follows [21]:

$$t_i = \frac{Q \times 3600}{m \times q} \ (\sec onds) \tag{3}$$

where t_i: irrigation duration in seconds, Q: irrigation dose (lt/bucket), m: number of drip lines per bucket and q: drip emission (lt/h) (Appendix A). We applied this formula as our baseline duration to determine our fuzzy set membership functions.

The rationale behind using FLC for controlling the pump stems from the use of an expert system based on linguistic if–then rules for the pump on/off duration, coupled with inherent uncertainty and errors in sensor readings and different parameters for different crops. The above constitute fuzzy modeling as an ideal fit for the proposed system. By abstracting the inputs and outputs using fuzzy sets, more robust control can be achieved.

Figure 11 shows the fuzzy membership functions. In order to maintain computational simplicity, we selected trapezoidal and triangular membership functions in the proposed control scheme to describe the linguistic values of the fuzzy input and output variables. The amount of overlap between the membership functions areas was chosen to have at most two membership functions overlapping; thus, we will never have more than six rules activated at a given time. This offers computational simplicity in the implementation of the proposed scheme, which was a design objective. Furthermore, it does not make intuitive sense to have a parameter such as pH being considered as both high and low according to hydroponics guidelines. The EC and pH membership functions were determined based on hydroponics guidelines, while the voltage was based on extending battery life. The irrigation duration membership functions were based on multiples of the baseline irrigation duration from Equation (3). In the case of clearly erroneous sensor values (Section 4.2), the values are not fed to the FIE; instead, the last valid values are used. If the pH and EC values seem correct they are fed to the FIE and the last valid values are updated for the next sensor reading.

The fuzzy rule base was determined based on hydroponics guidelines. As mentioned, three inputs are used, with three membership functions per input. Therefore, there are a total of $3^3 = 27$ rules. The rules were determined empirically according to the principle that when conditions are ideal, the irrigation period is the one determined by Equation (3). When conditions are suboptimal, especially when the input voltage is low, indicating the battery is discharged, the irrigation duration is shortened in order to conserve energy. When EC and pH indicate that there are few nutrients to be absorbed, then the irrigation period is also shortened. The forms of the rules with some indicative cases are shown in Table 4.

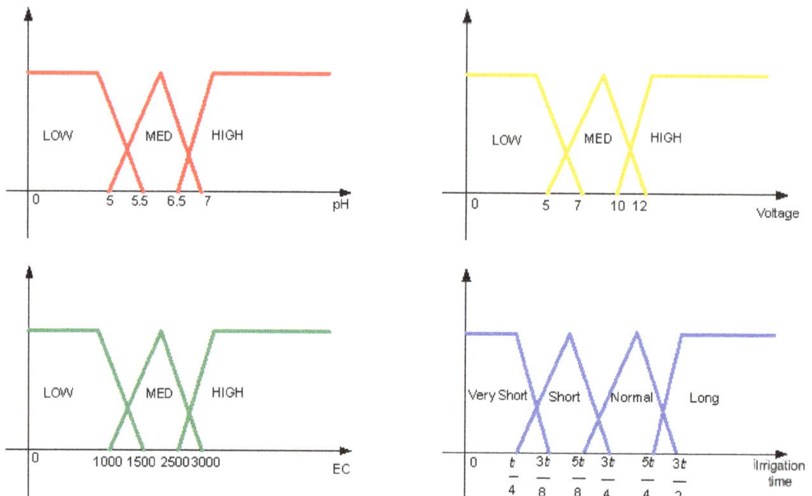

Figure 11. Fuzzy membership functions.

Table 4. Fuzzy rule structure with examples.

pH	EC	V	t_i
LOW	LOW	LOW	VERY SHORT
LOW	LOW	MEDIUM	SHORT
MEDIUM	LOW	HIGH	SHORT
MEDIUM	MEDIUM	HIGH	LONG
HIGH	HIGH	HIGH	SHORT

The execution time of the FIE was measured to be 3326 ms, which is included in the SCU main loop execution time reported in Section 3.1.

4. Deployment and Evaluation

4.1. System Deployment

The iPONICS system was operated in a lab environment for over a year (Figure 12) and was installed in a pilot hydroponics greenhouse in order to grow crops. The monitoring website is shown in Figure 13. The ubidots platform was used to upload the data, and our application was built using an API to transfer the data in .csv format and display them. The application was built using javascript, css and html. The user can monitor multiple greenhouses at a glance. It provides a dashboard for monitoring the water quality and a sidebar for alerts and warnings, such as sensor and SD card errors. The user can generate temperature, humidity, pH, EC, DO and water temperature graphs created using the D3.js library, with the default being the past 24 h. The water quality sensor readings are also shown as doughnut graphs that are color-coded the same way as the sensors (Figure 13) for ease of monitoring.

Figure 12. iPONICS system deployed in lab conditions.

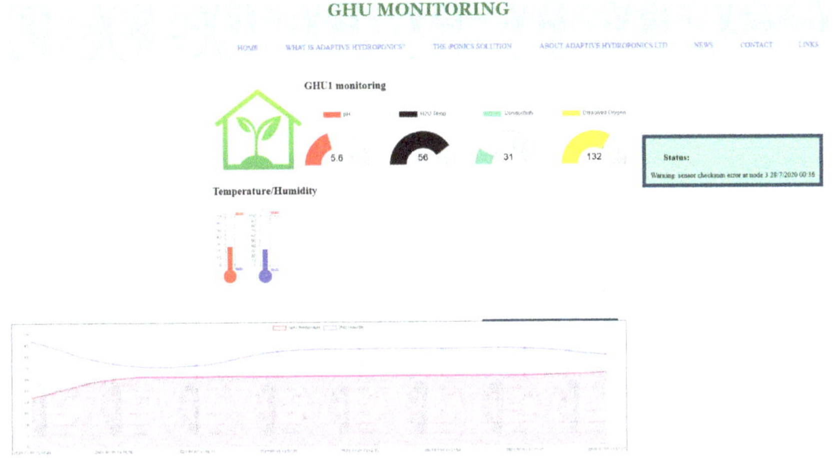

Figure 13. iPONICS monitoring page.

4.2. Preliminary Reliability Analysis

Since the above components are meant to operate continuously in greenhouse conditions for long periods in order to grow crops, we aimed to analyze their reliability at as early a stage as possible. We performed preliminary reliability analysis based on the errors and failures observed since they were operational, both in laboratory and greenhouse conditions. Reliability analysis was important in the iPONICS system because electronic failures could lead to crop failure, as well as a waste of energy and water.

We performed the analysis on the SCU and ESN nodes since the GSN nodes operated infrequently and were aimed at outside threats, which are irrelevant to the plant growth process. The central role of the SCU makes it the most critical unit in the subsystem. The ESN nodes have redundancy that allows fault tolerance, while there is a single SCU in each GHU. We had two SCUs and four ESNs operate continuously for close to 1000 h. Using that continuous operation as a basis, we observed errors and measured the mean time between them (MTBF). The errors are summarized in Table 5. They can be categorized as follows.

- Sensor errors—Generally, sensor errors can be divided into two main categories: hard failure (complete failure) and soft failure, such as bias, drift and outliers [27]. We observed no hard failures, but we observed both drift and outliers.

- Transmission errors—We noticed occasional GPRS transmission failures. Some were corrected by retransmission and we added automatic retries that generally solved the problem, but there was one occasion that required manual intervention by communicating with the internet service provider in order to be resolved.

Table 5. Error summary.

Unit	Error Type	MTBF	Redundancy	Recovery
SCU	Sensor error	3.5 days	Yes (on water temperature)	Automatic (usually on next measurement)
	Transmission failure (transient)	40 days	No	Automatic (retransmission)
	Transmission failure (requires intervention)	180 days	No	Manual
	Drift (pH sensor)	N/A	No	Manual (recalibration)
ESN	Sensor error	15 days	Yes	Automatic (usually on next measurement)

Since the system remains on all the time but sampling and transmission are done with a specific sampling rate, we used two approaches to model the system reliability. Specifically, we use a Poisson distribution assuming λ is equal to the MTBF:

$$P(n) = e^{-\lambda t} \frac{(\lambda t)^n}{n!} \tag{4}$$

Regarding sensor errors, we also use a binomial distribution with a failure probability equal to the error rate:

$$P(X = n) = \binom{N}{n} p^n (1-p)^{N-n} \tag{5}$$

We noticed that when the sampling rate was constant, the two distributions yielded identical results, which is to be expected as the Poisson distribution is a limiting case of a binomial distribution when the number of trials is large and the probability is small [28], which holds for our case (Table 6, column 2). We estimated the probability that there will be zero errors during a day at the highest sampling rate (no error events in all 24 samples), as well as the probability that there will be an error event in all samples in that day at a low sampling rate (4 samples/day). We were particularly interested in the case when there were errors in all pH and EC measurements since it is then impossible to monitor the system, as well as correctly decide on the irrigation duration.

Table 6. SCU error analysis.

Sensor Error	Error Probability	Probability of No Errors in One Day (24 Samples)	Probability of Only Errors in One Day (4 Samples)
Any single/multiple error	0.93%	80%	6.82211E-5
pH	0.40%	98.4%	2.51071E-6
EC	0.41%	98.3%	2.76579E-6
Transmission	0.1%	99.6%	1.04155E-6

As shown in the table above, the probability of single or multiple errors, including transmission errors, at the lowest sampling rate (worst case) was less than 7 in 100,000 or about 1 in 39 years, which is longer than the expected lifetime of the system. Furthermore, the probability that there will be only erroneous measurements in either pH or EC was less than the sum of respective probabilities since the events were not disjointed, which was only approximately 1 in 200,000.

5. Conclusions and Future Work

A novel hydroponics monitoring and control system based on IoT technologies was introduced. In particular, the system is composed of a specialized wireless sensor network for monitoring the essential parameters for hydroponics and control for the pump. It provides the greenhouse keeper with a user-friendly web-based tool to monitor their crops, as well as alarms and warnings, allowing for the observation of multiple greenhouses with minimal effort and need for intervention.

Our future work will focus on two directions. First, further reliability analysis by forcing errors and stress testing the system and, second, data analytics. Specifically, we are interested in predicting nutrient values based on the original concentrations and the water quality sensor values. In general, monitoring specific nutrients is a challenging problem for both terrestrial and space applications, even though several technologies exist [29]. Most of these technologies require costs that may be prohibitive to small-scale farmers, who often add nutrients empirically. This could lead to either plants being starved for nutrients or reaching toxic levels of nutrient absorption. For now, we are using specialized nutrient monitoring equipment [30] to obtain data points daily, as shown in Figure 14. Generally, EC is the aggregate of all ions in the plant nutrient solution. The nutrients are added to the water (day 1), increasing EC, and they are gradually absorbed, typically within 24 h. While the general trend is clearly visible in Figure 14, the individual nutrient absorption depends on the pH, the presence of competing nutrients, as well as other factors. The ion-specific electrode used requires frequent (daily) recalibration, limiting the ability to be used remotely. Accurately predicting the nutrient values only from the SCU readings would save the cost of such expensive equipment, as well as the need for recalibration, while protecting from possible empirical errors. It can also be used to lower the SCU sampling rate to conserve energy, even if it increases the probability of erroneous measurements. We are currently exploring our obtained data using machine learning models. At the time of this writing, given the low data rate since the nutrient values do not change rapidly and the fact that EC reflects the aggregate of all nutrients and is expected to be highly linear, we are exploring a simple regression model, but this is still under investigation.

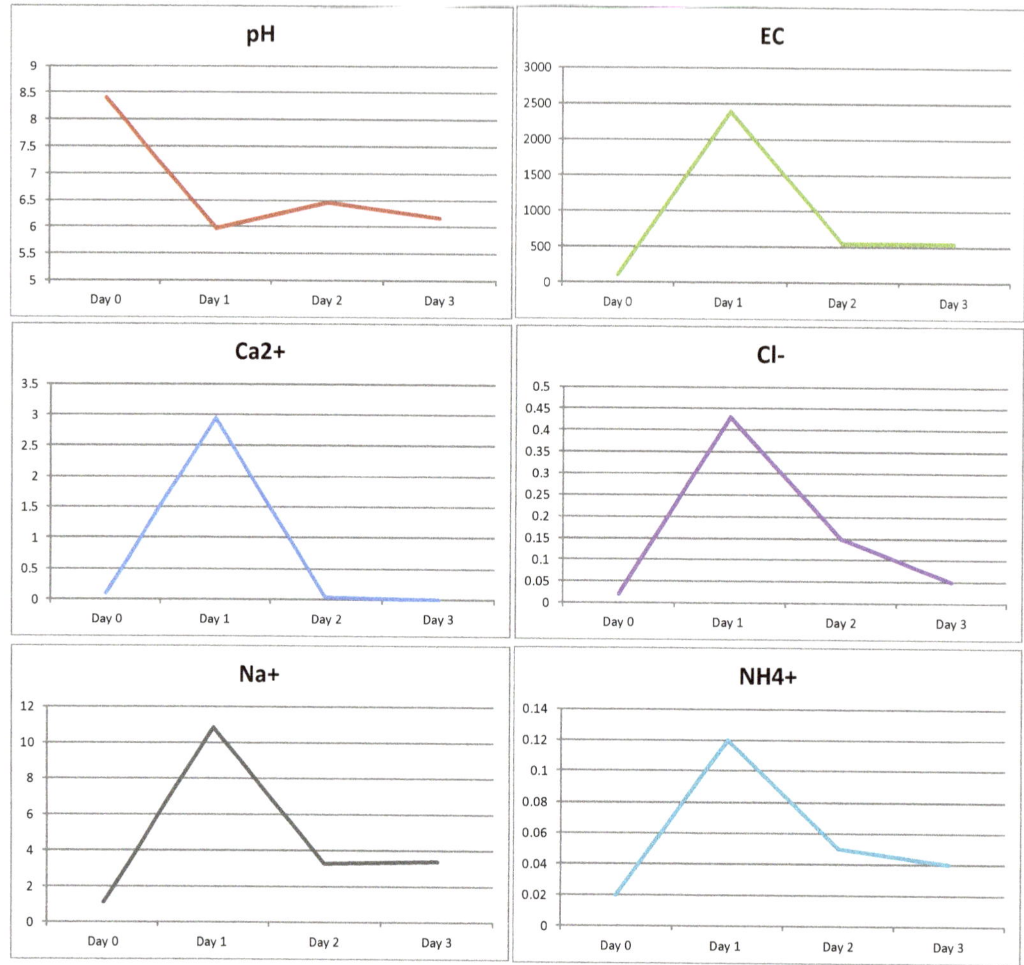

Figure 14. pH, EC and concentrations of various nutrients during a typical nutrient cycle: before nutrients are added or replenished (day 0) and after (day 1). Most nutrients are absorbed within one day (by day 2), depending on the pH.

Author Contributions: Conceptualization, K.T. and A.A.; Funding acquisition, K.T. and A.A.; Project administration, A.A.; Software, A.A.-Z. and C.Z.; Supervision, K.T.; Validation, A.A.-Z., N.C., M.C. and S.P.; Writing—review and editing, K.T. All authors have read and agreed to the published version of the manuscript.

Funding: This research was funded by the European Regional Development Fund and the Republic of Cyprus through the Research and Innovation Foundation (Project: START-UPS/ 0618/48 project title: "iPONICS: Smart Off-Grid System for Sustainable Hydroponics".

Institutional Review Board Statement: Not applicable.

Informed Consent Statement: Not applicable.

Data Availability Statement: Not applicable.

Conflicts of Interest: The authors declare no conflict of interest.

Abbreviations

The following abbreviations are used in this manuscript:

ICT	Information and communication technology
EC	Electrical conductivity
FIE	Fuzzy inference engine
FLC	Fuzzy logic control
MTBF	Mean time between failures
GHU	Greenhouse unit

Appendix A

Table A1. Mathematical symbols used.

Symbol	Meaning	Unit(s)
T	Sleep duration, roughly equal to sleep period	ms
t_e	Main loop execution time	ms
E	Energy consumption	mJ
t_i.	Irrigation duration	s
Q	irrigation dose	lt
m	Number of drip lines per bucket	
q	Drip emission	lt/h
λ	Error rate	errors/h
t	System operation time	h
n	Number of errors	
N	Number of samples	
P	Error/failure probability	

References

1. Peters, N.E.; Meybeck, M. Water Quality Degradation Effects on Freshwater Availability: Impacts of Human Activities. *Water Int.* **2000**, *25*, 185–193. [CrossRef]
2. Khatri, N.; Tyagi, S. Influences of natural and anthropogenic factors on surface and groundwater quality in rural and urban areas. *All Life* **2015**, *8*, 23–39. [CrossRef]
3. Atlas Scientific: Environmental Robotics. Available online: https://www.atlas-scientific.com/ (accessed on 28 November 2020).
4. Available online: http://www.libelium.com/products/waspmote/ (accessed on 28 November 2020).
5. Verdouw, C.; Wolfert, S.; Tekinerdogan, B. Internet of Things in agriculture. *CAB Rev. Perspect. Agric. Vet. Sci.* **2016**, *11*, 1–12. [CrossRef]
6. Tzounis, A.; Katsoulas, N.; Bartzanas, T.; Kittas, C. Internet of Things in agriculture, recent advances and future challenges. *Biosyst. Eng.* **2017**, *164*, 31–48. [CrossRef]
7. Wang, J.; Yang, Y.; Wang, T.; Sherratt, R.; Zhang, J. Big Data Service Architecture: A Survey. *J. Internet Technol.* **2020**, *21*, 393–405.
8. Wang, J.; Chen, W.; Wang, L.; Sherratt, R.; Alfarraj, O.; Tolba, A. Data Secure Storage Mechanism of Sensor Networks Based on Blockchain. *Comput. Mater. Contin.* **2020**, *65*, 2365–2384. [CrossRef]
9. Zhang, J.; Yang, K.; Xiang, L.; Luo, Y.; Xiong, B.; Tang, Q. A Self-Adaptive Regression-Based Multivariate Data Compression Scheme with Error Bound in Wireless Sensor Networks. *Int. J. Distrib. Sens. Networks* **2013**, *9*. [CrossRef]
10. Wang, Q.; Yang, C.; Wang, Y.; Wu, S. Application of low cost integrated navigation system in precision agriculture. *Intell. Autom. Soft Comput.* **2020**, *26*, 1433–1442. [CrossRef]
11. Liao, M.; Chen, S.; Chou, C.; Chen, H.; Yeh, S.; Chang, Y.; Jiang, J. On precisely relating the growth of Phalaenopsis leaves to greenhouse environmental factors by using an IoT-based monitoring system. *Comput. Electron. Agric.* **2017**, *136*, 125–139. [CrossRef]
12. Codeluppi, G.; Cilfone, A.; Davoli, L.; Ferrari, G. LoRaFarM: A LoRaWAN-Based Smart Farming Modular IoT Architecture. *Sensors* **2020**, *20*, 2028. [CrossRef] [PubMed]
13. Rajalakshmi, P.; Devi Mahalakshmi, S. IOT based crop-field monitoring and irrigation automation. In Proceedings of the 10th International Conference on Intelligent Systems and Control (ISCO), Coimbatore, India, 7–8 January 2016.
14. Danita, M.; Blessy, M.; Nithila, S.; Namrata, S.; Paul, J. IoT Based Automated Greenhouse Monitoring System. In Proceedings of the Second International Conference on Intelligent Computing and Control Systems (ICICCS), Madurai, India, 14–15 June 2018.

15. Sofwan, A.; Sumardi, S.; Ahmada, A.; Ibrahim, I.; Budiraharjo, K.; Karno, K. Smart Greetthings: Smart Greenhouse Based on Internet of Things for Environmental Engineering. In Proceedings of the International Conference on Smart Technology and Applications (ICoSTA), Surabaya, Indonesia, 20 February 2020.
16. Trilles, S.; Torres-Sospedra, J.; Belmonte, Ó.; Zarazaga-Soria, F.; González-Pérez, A.; Huerta, J. Development of an open sensorized platform in a smart agriculture context: A vineyard support system for monitoring mildew disease. *SUSCOM* **2020**, *28*, 100309. [CrossRef]
17. Tatas, K.; Chrysostomou, C. Hardware Implementation of Dynamic Fuzzy Logic Based Routing in Network-on-Chip. *MICPRO* **2017**, *52*, 80–88. [CrossRef]
18. Al-Zoubi, A.; Tatas, K.; Kyriacou, C. Fuzzy classification of OpenCL programs targeting heterogeneous systems. *JIFS* **2020**, *39*, 7189–7202. [CrossRef]
19. Al-Zoubi, A.; Tatas, K.; Kyriacou, C. Towards Dynamic Multi-task Schedulling of OpenCL Programs on Emerging CPU-GPU-FPGA Heterogeneous Platforms: A Fuzzy Logic Approach. In Proceedings of the 10th IEEE International Conference on Cloud Computing Technology and Science (CloudCom), Nicosia, Cyprus, 10–13 December 2018; pp. 247–250.
20. Douglas, J. *Advanced Guide to Hydroponics: (Soilless Cultivation)*, 2nd ed.; Pelham Books: London, UK, 1986.
21. Atlas Scientific, Lab Grade pH Probe. Available online: https://atlas-scientific.com/files/pH_probe.pdf (accessed on 28 November 2020).
22. Atlas Scientific, PT-1000 Temperature Probe. Available online: https://atlas-scientific.com/files/PT-1000-probe.pdf/ (accessed on 28 November 2020).
23. Atlas Scientific, Lab Grade D.O. Probe. Available online: https://atlas-scientific.com/files/LG_DO_probe.pdf/ (accessed on 28 November 2020).
24. Atlas Scientific, Conductivity Probe K 1.0. Available online: https://atlas-scientific.com/files/EC_K_1.0_probe.pdf/ (accessed on 28 November 2020).
25. Atlas Scientific, Tentacle Shield. Available online: https://atlas-scientific.com/tentacle-t1/ (accessed on 28 November 2020).
26. Ross, T.J. *Fuzzy Logic with Engineering Applications*, 4th ed.; John Wiley & Sons: Chichester, UK, 2016.
27. Alippi, C. Fault Diagnosis Systems. In *Intelligence of Embedded Systems*; Springer: Cham, Switzerland, 2014.
28. Pishro-Nik, H. *Introduction to Probability, Statistics, and Random Processes*; Kappa Research LLC: Montgomery, PA, USA, 2014. Available online: https://www.probabilitycourse.com (accessed on 28 November 2020).
29. Bamsey, M.; Graham, T.; Thompson, C.; Berinstain, A.; Scott, A.; Dixon, M. Ion-Specific Nutrient Management in Closed Systems: The Necessity for Ion-Selective Sensors in Terrestrial and Space-Based Agriculture and Water Management Systems. *Sensors* **2012**, *12*, 13349–13392. [CrossRef] [PubMed]
30. CleanGrow Nutrients. Available online: https://www.ionselectiveelectrode.com/pages/userguides (accessed on 28 November 2020).

 *technologies*

Article

Analysis of the Impact of Electrical and Timing Masking on Soft Error Rate Estimation in VLSI Circuits [†]

Pelopidas Tsoumanis [1,*,‡], **Georgios Ioannis Paliaroutis** [1,*,‡], **Nestor Evmorfopoulos** [1] and **George Stamoulis** [1,2]

1   Department of Electrical and Computer Engineering, University of Thessaly, 38334 Volos, Greece; nestevmo@e-ce.uth.gr (N.E.); georges@e-ce.uth.gr (G.S.)
2   Department of Computer Science, University of Thessaly, 35131 Lamia, Greece
*   Correspondence: petsouma@e-ce.uth.gr (P.T.); gepaliar@e-ce.uth.gr (G.I.P.)
†   This paper is an extended version of our paper published in Proceedings of the 34th IEEE International Symposium on Defect and Fault Tolerance in VLSI and Nanotechnology Systems (DFT), Athens, Greece, 6–8 October 2021.
‡   These authors contributed equally to this work.

Abstract: Due to continuous CMOS technology downscaling, Integrated Circuits (ICs) have become more susceptible to radiation-induced hazards such as soft errors. Thus, to design radiation-hardened and reliable ICs, the Soft Error Rate (SER) estimation constitutes an essential procedure. An accurate SER evaluation is provided based on a SPICE-oriented electrical masking analysis, combined with a TCAD characterization process. Furthermore, the proposed work analyzes the effect of a Static Timing Analysis (STA) methodology and the actual interconnection delay on SER evaluation. An analysis of the generated Single Event Multiple Transients (SEMTs) and the circuit operating frequency that are related to the SER estimation is also discussed. Various benchmarks, synthesized utilizing a 45 nm and 15 nm technology, are employed, and the experimental results demonstrate the SER variation as the device node scales down.

Keywords: electrical masking; interconnection delay; Single Event Multiple Transients; Soft Error Rate; STA; TCAD; timing-masking; transient faults

Citation: Tsoumanis, P.; Paliaroutis, G.I.; Evmorfopoulos, N.; Stamoulis, G. Analysis of the Impact of Electrical and Timing Masking on Soft Error Rate Estimation in VLSI Circuits. *Technologies* **2022**, *10*, 23. https://doi.org/10.3390/technologies10010023

Academic Editors: Spiros Nikolaidis and Rodrigo Picos

Received: 29 December 2021
Accepted: 28 January 2022
Published: 31 January 2022

Publisher's Note: MDPI stays neutral with regard to jurisdictional claims in published maps and institutional affiliations.

1. Introduction

Over the past decades, the CMOS technology downscaling trend, which involves—among other factors—the shrinking of transistor dimensions, the reduction of supply voltages, and the increase in operating frequencies, has rendered modern Integrated Circuits (ICs) substantially susceptible to radiation-induced Single Event Transients (SETs) [1]. Thus, ICs' reliability regarding Soft Errors constitutes a challenging field of research, especially when it concerns critical systems. A radiation particle of sufficient energy that strikes on a gate may generate, under certain circumstances, a glitch at the output. At the circuit level, a soft error emerges when a SET is captured by at least one storage element. While the soft errors are not permanent, they could potentially pose a severe threat to vulnerable modern chips, especially critical ones. A factor that aggravates this problem is the emergence of Single Event Multiple Transients (SEMTs) that, primarily due to the increase in the circuit device density, as a consequence of the Moore's law, have become more prevalent recently [2]. In light of the above, the accurate evaluation of the Soft Error Rate (SER) constitutes a vital process to determine the ICs susceptibility to radiation hazards. Usually, the SER is measured in terms of FIT (Failures In Time), which is a widely-used reliable metric across different SER estimation tools for the assessment of chips' reliability.

To accurately estimate the SER of a design, it is indispensable to model sufficiently and accurately the three mechanisms that are able to impede an SET from propagating through a circuit and, eventually, being latched by the flip-flops (FFs), thus producing a soft error. These effects are logical, electrical, and timing-masking [3]. The logical masking occurs when an SET is masked on a subsequent gate because one of the other inputs is in a controlling value. For example, the controlling value of an AND gate is logic 0, whereas logic 1 is the controlling value of an OR gate. The electrical masking is associated with the electrical properties of the gates and the SET pulse characteristics and occurs as the SET pulse attenuates during the propagation through the logic gates, becoming too small to be reliably latched at the FFs. Finally, the timing-masking occurs when an SET arrives outside the latching window of the FF as this is determined by setup and hold times.

The modeling of the three masking mechanisms is equally significant in obtaining an accurate estimation of a circuit's vulnerability to radiation hazards. The logical masking is quite straightforward to model, and there are no major differences among the various SER estimation works, with the more prevalent model being the utilization of a simple logic level simulator. However, the electrical and timing-maskings can be modeled in many different ways, which determine the overall accuracy of the SER estimation. Plenty of work that attempts to model accurately the electrical masking mechanism has been proposed so far. Generally, there are two types of approaches that dominate the bibliography. The first is based on SPICE simulations to accurately characterize the propagated SET pulses and form LookUp Tables (LUTs), whereas the other attempts to estimate the effect of electrical masking through analytical modeling. The advantage of the former is its accuracy, even though it is expensive in terms of time, whereas the latter is more efficient but lacks accuracy. A closed-form approximation of the logic level waveforms, induced by α-particles on inverters considering the transistor's pull-up and pull-down network, the particle charge, and the capacitive loads, is presented in [4]. In [5], a closed-form expression is introduced to calculate the output voltage (amplitude) of a propagated transient fault. However, in [6] a simple ramp approximation equation is used to estimate the SET pulse width at the gate output. An approach that utilizes discrete values of the input waveform to approximate the whole output pulse (amplitude and duration) is presented in [7]. In [8], a pre-characterization library process, based on the SET pulse height and width at the gate input, is carrried out to extract parameters and form simple analytical continuous functions for pulse propagation. In [9], a two-phase pre-characterization process is performed with SPICE, forming LUTs utilized in the extraction of SET propagation mathematical equations. In [10,11], SPICE simulations are performed to characterize the SET width. Other approaches use transistor models to effectively model the electrical masking effect [12,13]. Recent works have revealed another significant aspect of SET propagation that affects its pulse width. In particular, the investigation of the SETs production and propagation in CMOS logic circuits, in [14], has shown that they may propagate without attenuation when they are generated from particle strikes of certain linear energy transfer. In [15], the authors present the Propagation-Induced Pulse-Broadening (PIPB) effect that a SET is subjected to as it propagates through long inverter chains. A direct relationship is reported between the SET pulse attenuation or broadening with the circuit design parameters and the gate delay [16,17]. The TCAD simulations of inverter chains in [18] relate the PIPB effect to the transistor voltage threshold. The impact of the propagation paths (including reconvergent paths), the input patterns, and the polarity of the SET on the PIPB effect is presented in [19]. The authors in [20] characterize the PIPB effect on SETs generated from a heavy ion microbeam and associate the transistor size with its confinement. All these works indicate that this aspect should be taken into account to achieve an accurate SER evaluation. A similar phenomenon of SET pulse broadening after narrowing (PBAN) due to the charge sharing is examined in [21]. As regards the timing-masking modeling, there is a lack of information in the bibliography upon the impact of each timing analysis approach on the SER estimation. A modified static timing analysis for the timing-masking modeling is proposed in [22], but the utilized delay model is not clarified, and the interconnection delay is ignored, which may underestimate or overestimate the results.

This work aims at placing an emphasis on the significance of the electrical and timing-masking model that is utilized from a SER estimation tool to achieve an efficient and accurate evaluation of modern ICs' vulnerability to radiation hazards. The applied timing analysis holds a key role in the modeling of both electrical and timing-masking mechanisms. Based on the results of the STA analysis, with respect to the fall/rise delays and the SPICE simulations, we are able to determine the SET pulse width as it propagates through the logic gates. Since the output pulse width depends on the input of the SET, which implies different fall and rise delays, the modeling of electrical masking becomes dynamic. As regards the timing-masking, the STA analysis that we implement ensures that the delay of the SET is calculated accurately as it propagates until FFs. The impact of interconnection wiring on SET pulse propagation delay is also discussed. The results of the electrical and timing-masking are validated with SPICE, indicating fairly good accuracy. Some preliminary results of this paper were presented in [23].

The rest of this work is organized as follows. Section 2 presents a TCAD simulation-based approach to achieve an accurate characterization of the radiation-induced SET pulses. Section 3 elaborates on the analysis and modeling of electrical masking and discusses the relationship of this model with the timing analysis approach selected for the SER estimation. Section 4 briefly highlights the main steps of the integrated SER evaluation flow that the aforementioned masking models are incorporated into. Section 5 verifies the electrical and timing-masking models and presents a series of experimental results on a variety of benchmarks regarding the impact of these models on SER estimation. Finally, Section 6 concludes this work.

2. Radiation-Aware TCAD Simulations

Radiation-induced soft errors constitute a reliability issue that is related to physics phenomena, which supersede the particle strike event, and which involve the interaction between the high-energy particle that strikes the silicon and the particles of the device. A certain approach to model such phenomena is real-time experiments through neutron beam setups and actual measurements of DUTs. However, they are time consuming and thus inefficient to model and characterize the electrical behavior of the devices across a wide spectrum of technology nodes. Therefore, TCAD tools are extremely useful in modeling, simulating, and optimizing the semiconductor process technologies and devices since they provide solutions that are based on deep physical equations.

One of the features of TCAD tools is that they provide an effective method to model and simulate the impact of SETs on semiconductor devices through their integrated physics background. In this work, various TCAD simulations are performed to identify the pulse of the generated SET. In particular, several simulations of particles striking the drain of NMOS and PMOS transistors model the induced disturbance. Figure 1 shows the charge generation as a heavy ion strikes the drain of an NMOS transistor with 90° angle, which is vertical to the drain contact, and its distribution within the transistor's mesh. Additionally, note that the width and the length of the NMOS drain are $W = 25$ nm and $L_d = 10$ nm, respectively.

The energy of the radiation particle corresponds to the *Linear Energy Transfer* (LET) that delivers as it penetrates the semiconductor and is a crucial factor in such type of simulations. The resultant charge generation as a heavy ion strikes the drain is reflected in the drain current. Figure 2 presents the drain current for three different LETs of the heavy ion. This graph shows that the current spike elevates as the energy increases, indicating the significance of heavy ion severity in the generation of SETs and, subsequently, in the emergence of soft errors.

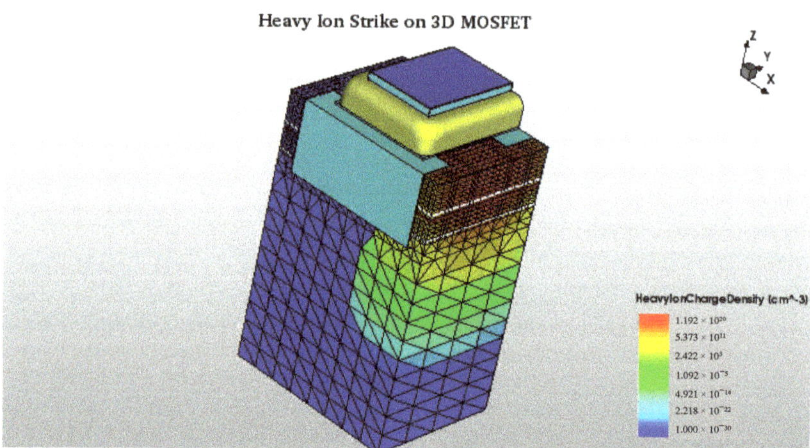

Figure 1. A heavy ion striking the drain of NMOS and inducing charge generation.

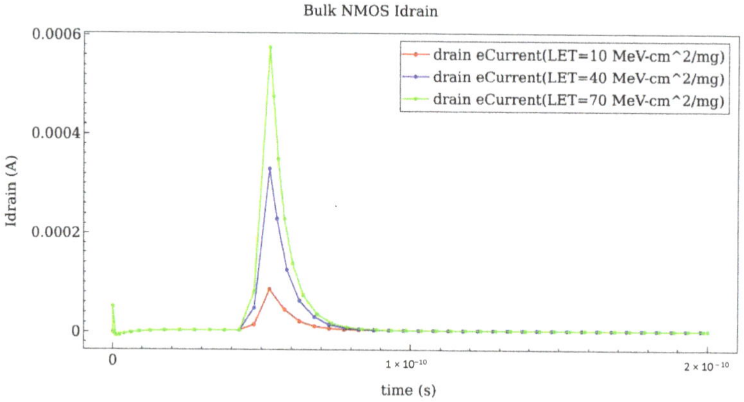

Figure 2. Drain current of NMOS for different LET values of the heavy ion.

A radiation-aware simulation of an NMOS or PMOS device might be sufficient to model the impact of such hazards on semiconductor devices, but at a logical level, there is an additional step that should to be considered. That is, we need to examine if the charge generation within the device emerges at the output of the corresponding logic gate and, if so, identify the shape (i.e., height, width, and slew) of the SET pulse. Thus, a mixed-mode simulation that combines the semiconductor devices based on physical models (TCAD) with devices described with compact models (SPICE) can be performed to observe the generated SET. In this way, the simulation of a heavy ion striking a node becomes more accurate as we are able to incorporate into the simulation the physical devices (i.e., transistors) that are modeled with a powerful TCAD tool. Figure 3 demonstrates a transient analysis of an inverter, performed with mixed-mode simulation, when two heavy ions strike the gate at different time moments. Note that the time moments are quite close, considering such a small time frame, which a rather unlikely scenario for two radiation particles striking the same cell within a few picoseconds. Therefore, this simulation setup is deliberately presented in this mode, to exhibit in the same graph the effect of SET pulse generation at the output of a logic cell in two cases. The first considers a particle striking the off PMOS transistor device and the second a particle striking the off NMOS, generating at the output a small and a large glitch, respectively. The difference in the SET amplitudes is attributed to the higher sensitivity of NMOS transistor compared to PMOS.

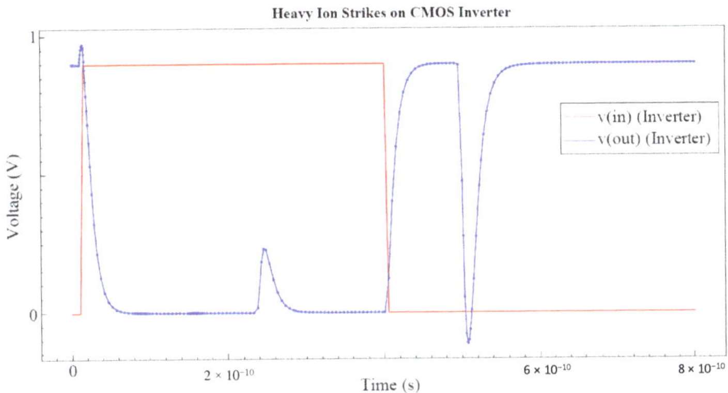

Figure 3. Transient plot of a CMOS Inverter when two heavy ions strike the NMOS and PMOS transistor at different time moments.

3. Electrical and Timing-Masking Model

The most significant factor in the estimation of SER in the combinational logic of ICs is the modeling of the three masking effects. Since the masking mechanisms are contingent on the connectivity and the design properties of the individual circuits, their impact varies from one circuit to another. Figure 4 presents the distribution of the generated SETs in an SER estimation process with respect to the type of masking (i.e., logical, electrical, and timing) that occurs or not when the same number of faults are injected. While logical masking appears to be the most prevalent factor contributing to the SET elimination, especially for the large-scale benchmarks, electrical and timing-masking jointly tend to be similarly substantial, if not more. This graph reveals the necessity of a sufficient and accurate modeling of masking mechanisms since this reflects on the SER estimation eventually.

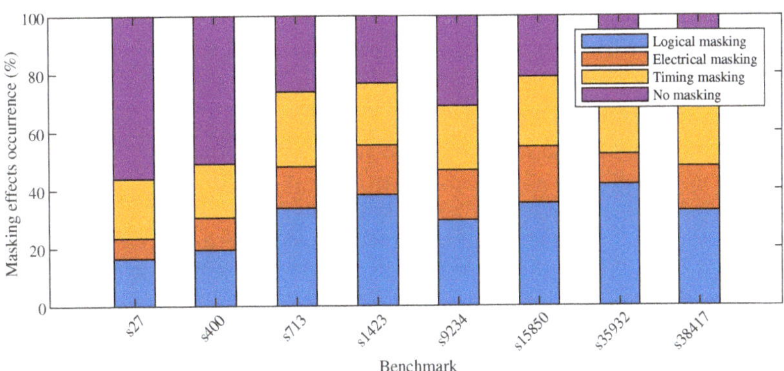

Figure 4. Impact of the masking effects on the propagation of radiation-induced SETs through some benchmark circuits for 45 nm technology.

3.1. Electrical Masking

The propagation of SETs is affected, to a great extent, by electrical masking, which is a critical factor that should be modeled sufficiently. In the context of this work, a comprehensive modeling of electrical masking that contributes to the SER evaluation accuracy is provided. SPICE simulations constitute a widely-utilized practice to model the SET pulse generation. In order to analyze gate susceptibility at the transistor level, current pulses are connected on the transistor nodes to generate SETs at the gate output. Therefore, to model the disturbance induced by the particle strike, which results in a 1→0 or 0→1 momentarily transition, current pulses are injected into PMOS and NMOS transistors. Generally, it should be highlighted that the modeling of the SET pulse is a crucial procedure since it may propagate through the circuit and be latched by a storage element if it is of sufficient duration and amplitude.

Figure 5 shows a SET pulse, which passes through a CMOS logic gate (e.g., an inverter) and is modeled as a trapezoidal waveform. The transition from logic 1 (high voltage) to logic 0 (low voltage) and from logic 0 to logic 1 constitutes the individual propagation delays for the rise and fall transitions of the output pulse, respectively, and is employed to determine the pulse propagation delay. More specifically, the high-to-low propagation delay t_{HL} is the time interval from the point that input reaches 50% of the voltage supply (VDD) (i.e., logic 0 to logic 1 transition) to the point that output reaches 50% of VDD (i.e., logic 1 to logic 0 transition). The low-to-high propagation delay t_{LH} is determined similarly.

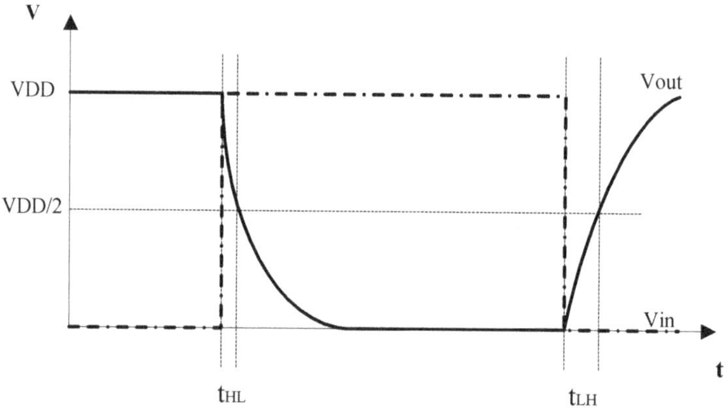

Figure 5. A SET pulse at the input and output of an Inverter and its HL and LH propagation delays.

Through an adequate number of SPICE simulations, it is observed that the transient pulse is deformed as it propagates through a logic gate, thus allowing for the modeling of the SET pulse propagation through a logic path. SET pulses of various output capacitance loads and several widths are taken into consideration. It should be emphasized that the applied output capacitance corresponds to the number of fanouts that a gate may have as well as the interconnection parasitics at its output, which means that it is a significant parameter for the delay of the pulse.

Table 1 presents the results of the SPICE simulations on the NAND2 and NOR2 gates when a SET pulse emerges on one input, whereas the other input is at a non-controlling value, so as the pulse is not logically masked, resulting in a 0→1→0 transition. The t_{LH} and t_{HL} as well as the output pulse width are measured for the different values of the output capacitance, indicating that they are directly related. It is clearly noticeable that the output pulse width results from the propagation delays t_{LH} and t_{HL} and, particularly, their difference. Therefore, for NAND2 gate the output pulse is calculated using Equation (1), whereas for NOR2 gate the output pulse is obtained from Equation (2).

Table 1. Propagation delays and output pulse widths for the 0→1→0 transition.

	V_{in} (mV)	Capacitance								
		1fF			5fF			10fF		
		t_{HL}	t_{LH}	V_o	t_{HL}	t_{LH}	V_o	t_{HL}	t_{LH}	V_o
NAND2	100	26.6	13.4	114.2	76.3	45.1	132.2	-	-	0
	300	26.4	13.4	313.9	86.6	45.1	342.5	163.8	84.3	380.5
	500	26.3	13.4	513.9	88.3	45.1	544.1	164.4	84.3	581.1
NOR2	100	37.1	10.5	74.3	-	-	0	-	-	0
	300	37.1	11.4	275.2	124.8	32.1	208.3	236.1	19.4	84.3
	500	37.1	11.4	475.3	124.8	37.3	413.5	236.2	54.5	318.8

Similarly, Table 2 presents the respective pulse characteristics for the opposite transition, i.e., 1→0→1. However, compared to the previous case, the pulse width of NAND2 gate is calculated with Equation (2).

$$V_o = V_{in} + (t_{LH} - t_{HL}) \tag{1}$$

$$V_o = V_{in} + (t_{HL} - t_{LH}) \tag{2}$$

Table 2. Propagation delays and output pulse widths for the 1→0→1 transition.

	V_{in} (mV)	Capacitance								
		1fF			5fF			10fF		
		t_{HL}	t_{LH}	V_o	t_{HL}	t_{LH}	V_o	t_{HL}	t_{LH}	V_o
NAND2	100	13.5	26.7	87.8	-	-	0	-	-	0
	300	13.8	26.7	288.1	43.3	86.9	257.4	51.4	164.8	187.6
	500	13.8	26.7	488.1	44.1	86.9	458.2	78.4	164.8	414.6
NOR2	100	11.5	36.9	126.4	37.8	119.1	182.3	70.4	99.5	130
	300	11.5	37	326.5	37.8	128.9	392.1	70.4	235.7	466.3
	500	11.5	37.2	526.7	37.8	125.1	588.3	70.4	235.9	665.7

The main observation from the SPICE results is that when a pulse propagates through a gate, its width may broaden or attenuate, depending on the transition and the gate type. Furthermore, note that for the 0→1→0 transition, as presented in Table 1, and for high capacitance values, the output pulses are equal to zero when SET width is 100 ps. This is due to the fact that the amplitude of the particular output pulses does not exceed the transition threshold (i.e., VDD/2), which means that it is not sufficient to propagate to the next stage and settle to a faulty voltage level. Additionally, it is worth to mention that there is a slight divergence between the measured output pulse and the actual difference between t_{HL} and t_{LH} delays, which results from the SPICE simulations and the parameters of the transistor models utilized.

Generally, there are difficulties in the practice of creating LUTs from SPICE simulations to model electrical masking, even though it is considered accurate. In particular, it is unfeasible to consider and analyze all the possible SET pulses that may emerge in a circuit since their shape characteristics change continuously when propagating through numerous circuit paths. Besides, complex LUTs may arise from the number of different factors that should be considered, such as the slews of the SET pulse, the parasitic capacitance, and the type of gates, thereby increasing the calculation cost and the memory usage. At the same time, this characterization procedure needs to be conducted for each utilized CMOS technology. These severe shortcomings are bypassed in this work by implementing an enhanced timing analysis methodology. Based on the deduction that the pulse propagation is directly related to the propagation delays t_{LH} and t_{HL}, the output pulse width is calculated considering the transition of the pulse and utilizing the corresponding equation. The propagation delays are computed once, during the basic STA, rendering the electrical masking accurate and fast, compared to the expensive LUT-based approaches.

3.2. Timing Masking

The accurate analysis of timing-masking is vital in the SER estimation of a circuit since the timing properties of the SET when arriving at the FF input and the timing circuit parameters affect the emergence of a soft error. Therefore, for the circuit timing behavior investigation, i.e., the determination of both the gates delay and the critical path, utilizing a basic STA methodology is significant. In particular, the STA method is based on LUTs, which store the input transition rates and load capacitances for each logic cell. The LUTs are obtained from the properly defined Non-Linear Delay Model (NLDM) of CMOS libraries and are formed under typical, worst, fast, and slow case conditions. Thus, accurate modeling of the timing-masking can be achieved when this is based on an accurate STA methodology.

At the early stages of SER evaluation and disregarding gate logic values, the circuit critical path estimation is conducted by implementing the STA method. In this analysis, based on the timing sense of gate input pins, the propagation delay of the gate is calculated by taking into consideration the maximum delay of the individual input arcs. At the later stages of electrical and timing-masking, this analysis can be enhanced though. In particular, the SET propagation delay when passing through a gate, which is needed for the timing-masking modeling, is obtained by observing its transition and the input that emerges. Finally, we achieve a result that approximates the SPICE simulation results, by taking into account the actual propagation delay for the particular input instead of the maximum delay among all the inputs. As a result, in the context of timing-masking modeling, the enhanced STA is converted, in a sense, into a Dynamic Timing Analysis (DTA). Additionally, note that this analysis is made only for forward logic cones, i.e., the logic paths that the SET propagates until the FF inputs.

3.3. Interconnection Delay

The performance of the modern CMOS ICs, in terms of operating frequency, power consumption, etc., is to some extent affected by the interconnect wiring among their logic components (e.g., cells, blocks, etc.). The wiring within a circuit introduces parasitic quantities of resistance (R), capacitance (C), and inductance (L) that, jointly with the logic gates, determine the propagation delay of the signals. There are various approximate techniques that model and estimate the interconnection delay, during the pre-layout phase, by taking into account the gate fanouts and estimating the total wire length for the entire circuit. However, the actual interconnection network of a design can be obtained only after the Placement and Routing (P&R) process with the extraction of the corresponding Standard Parasitic Exchange Format (SPEF) file. This file represents the parasitic wire data, which include parasitic resistance, capacitance, and their interconnection, and may be further used for simulation purposes such as delay calculation.

Figure 6 presents a typical example of an existing RC network of a net as extracted from the SPEF file of a design. That is a distributed net model and is depicted as an RC-tree with two branches, which is the number of fanouts. Given such an RC network, we can compute the interconnection delay by applying various models, such as the traditional Elmore's delay model [24]. However, its calculation is not presented since it is out of the scope of this work.

In the context of SER estimation, it is important to take into consideration the impact of interconnection delay on the SET pulse propagation. As regards the incorporation of the interconnection network into the SER estimation tool, a SPEF file parser has been implemented to account for the parasitics of each net and, then, estimate their delay. Moreover, the pulse width at the output of a gate is transformed to a new one at the inputs of the fanout gates considering the current parameters, such as slew and total wire capacitance. Thus, detailed modeling of the interconnection network is accounted for to obtain even more accurate results regarding circuit's susceptibility.

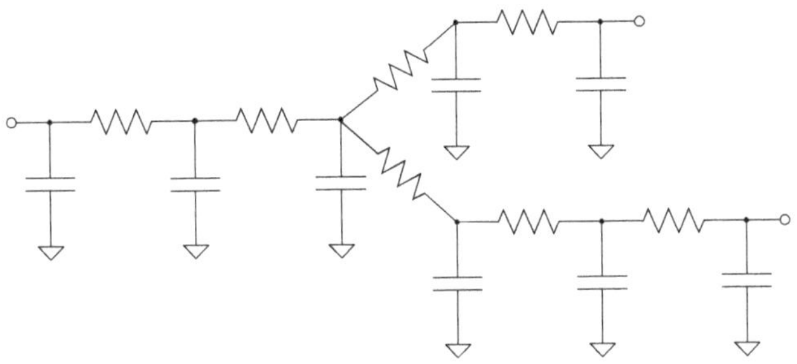

Figure 6. Distributed RC interconnection tree.

4. SER Evaluation Integrated Process

To evaluate the SER considering the proposed electrical and timing-masking models, the ISCAS '89 benchmarks were synthesized with Synopsys® Design Compiler™, using the 45 nm and 15 nm Open-Cell libraries [25], and their layouts were extracted using Cadence® Innovus™EDA tool. Figure 7 describes the overall procedure of this work.

In the previous section, we mentioned that the interconnection delay is calculated utilizing the SPEF files, which is an input to this tool. Additionally, the identification of the gate transistors position and the sensitive regions of each gate on the die, along with the circuit creation, are accomplished through the parsing of the DEF and GDSII files, which are extracted during the P&R process for the corresponding benchmarks. A gate's output may be changed only if its sensitive regions are affected by a particle strike. SPICE simulations for both NMOS and PMOS transistors for all input combinations are performed by injecting current pulses extracted from Synopsys® Sentaurus™TCAD tool for different LETs and observing the output pulse to identify gate sensitivity. This process shows that, according to current input values, sensitive regions are regarded as inactive transistors, that is, NMOS and/or PMOS diffusion.

An integrated tool based on Monte-Carlo simulations, modeling the three masking phenomena (logical, electrical, and timing) that affect the probability that a transient fault will become a soft error and emphasizing on SEMTs analysis, is utilized for the SER evaluation process [26]. An SEMT occurs when a heavy ion strikes a sensitive area over the chip, producing glitches on adjacent cells. To achieve a detailed evaluation of the circuit sites sensitivity to radiation, each circuit layout is divided into several grids. The execution time is reduced considerably by exploiting this idea and implementing a parallel SER estimation of the grids. Two other significant parameters are the handling of reconvergent pulses and the effect of temperature. The former is something that is taken into account since the transient faults following multiple paths and reconverging at a subsequent gate are not negligible, whereas the latter is also considered, indicating that SER becomes greater when the temperature increases since the SET widths become more intense [26]. Furthermore, the electrical and timing-masking models, as they are described in the manuscript, depend mainly on the technology library utilized (45 nm and 15 nm standard cell library). In particular, the critical factor is the timing model that we choose from the current library. So, even if it seems inconvenient to perform such simulations for different technologies, it is attainable since one of our tool's inputs is the technology library files. Finally, in this work we focus mainly on radiation as the primary source of SETs since it is the usual cause of glitches. However, our method can be readily expanded to include other sources of soft errors by considering the relevant induced charges (for example, currents delivered at the circuit inputs via mechanisms such as latchup).

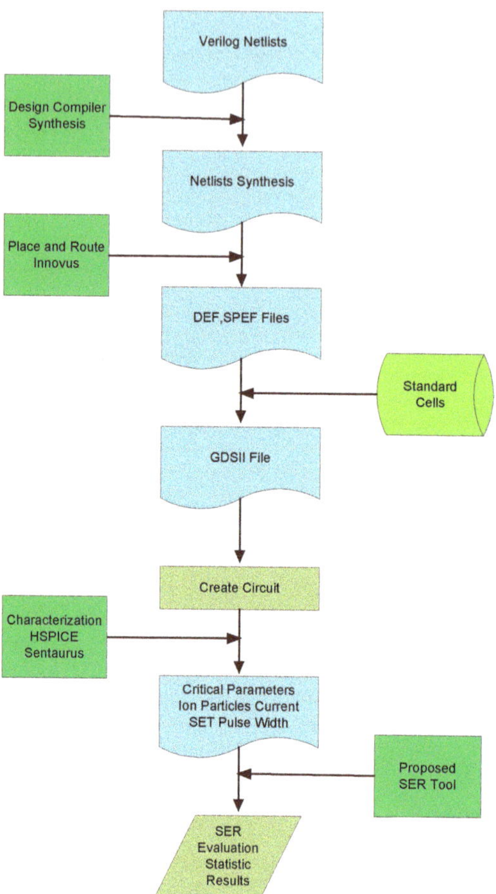

Figure 7. Overall SER simulation process.

5. Experimental Results

In this section, we provide a verification framework for the electrical and timing-masking modeling and demonstrate various experimental results regarding some critical factors in SER estimation and the different electrical and timing approaches. All the experiments are performed on an Intel Core i7-4790 @3.60 GHz machine with a Linux-based OS and 16 GB of RAM, whereas some of the ISCAS '89 benchmark designs are utilized to evaluate this work and demonstrate the results. Additionally, the SER is estimated both in terms of FIT and as a probability.

5.1. Electrical and Timing Verification

To verify both the electrical and timing-masking that are implemented from the SER estimation tool, we extract different logic paths with respect to the number and type of logic gates from various benchmark designs. Subsequently, SPICE simulations are performed to model the propagation of SETs. More specifically, each path, which is a circuit part, is imported to SPICE, and a SET pulse is applied on the input of the first gate to perform a simulation and obtain the pulse width and overall path delay at the output of the last gate. The first two paths are extracted from the s27 design and the others from s298 and s400 designs. Furthermore, note that the other gate inputs are in non-controlling value during the simulation to impede logical masking occurs. Besides, we use the SER estimation tool to simulate the selected paths, applying a SET of the same width with SPICE to observe the pulse shape at the end of the path. The checked paths, the length of each one, and the output SET pulse width and propagation delay for both SPICE and tool's simulations are shown in Table 3. The accuracy of the proposed approach reaches about 96%, which is acceptable considering the difference in execution time, since SPICE simulation is considered time consuming. Finally, note that this comparison is sufficient for the verification of the SER methodology as well, since the electrical and timing-maskings are two of the most important factors in SER estimation.

Table 3. Comparison of the proposed electrical and timing-masking models with SPICE on SET pulse propagation paths.

Path	Gate Stages	Electrical			Timing		
		Spice	Tool	Acc.	Spice	Tool	Acc.
1	5	201	211	95%	97	101	96%
2	8	199	188	94.5%	110	118	93%
3	13	202	214	94%	210	219	96%
4	20	197	185	94%	330	344	96%
5	25	194	186	96%	420	446	94%
6	36	203	217	93%	607	649	93%

5.2. Effect of SEMTs and Operating Frequency on SER Estimation

As discussed in Section 4, the overall circuit SER is obtained by modeling the three masking mechanisms during the transient fault propagation, at logic level. The area that SEMTs may occur is determined by the energy of the radiation particle strike, whereas the shape (i.e., length, width, and slope) of the generated glitches depends mainly on the load capacitance and the device characteristics of each gate, e.g., the channel length, the width of diffusions. A logic cell characterization of SET pulse widths was conducted with TCAD and mixed-mode simulations to facilitate an accurate SER estimation. The TCAD simulations considering a conventional 45 nm MOSFET planar technology and a 15 nm FinFET non-planar (3D) technology indicated that the latter is more resistant to heavy ions than the former.

In particular, Figure 8 presents the generated SET pulse widths of an inverter —for these technologies—when heavy ions of different LETs strike the cell. We observe that while the LET of the heavy ion increases, the SET pulse width increases as well for both process technologies. However, the widths for the 45 nm increase steeply and are much greater compared to those for the 15 nm, which shows that the FinFET technology is more resistant to ionizing radiation. Additionally, something that should be underlined is that this difference is primarily attributed to the gate transistor layout and not to the different technology nodes.

The downscaling of the node technology comes with higher operating frequencies. However, in an SER point of view when the frequency increases, the probability that an SET will be latched by a memory element increases as well. Table 4 presents the minimum clock period for some benchmarks as this was calculated during the timing analysis. The period of the benchmarks synthesized with respect to the 15 nm technology is significantly decreased compared to the period of the benchmarks at 45 nm technology. This means that it is more probable that an SET will not be masked, which is expected to reflect on the SER.

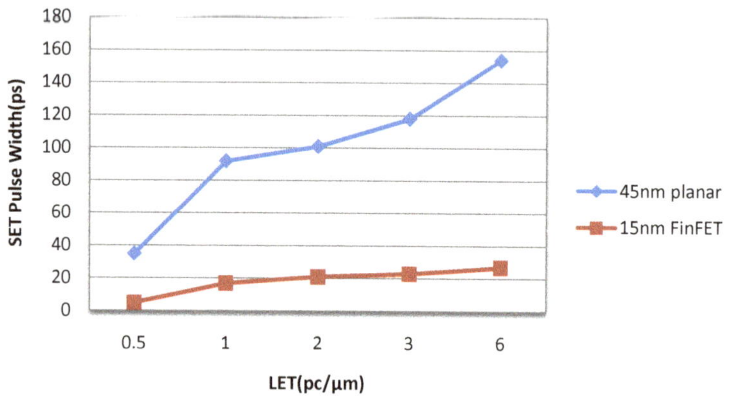

Figure 8. SET pulse width of Inverter as a function of LET for 45 nm planar and 15 nm FinFET technologies.

Table 4. Calculated clock period of some benchmarks for 45 nm and 15 nm technologies.

Benchmark	Clock Period (ps)	
	45 nm	15 nm
s298	642	75
s400	581	92
s5378	994	124
s15850	3098	290
s35932	10,156	960

Another critical aspect of SER estimation that is affected from the downscaling of the device node is the number of the SEMTs that may appear in a simulation. An analysis of the circuit sites that are affected from the injection of radiation particles on different locations of the circuit layout is presented in the following tables. In particular, Table 5 presents the total number of particles injected, the number of SEMTs that these particles created, the overall number of affected gates from the corresponding SEMTs, and (finally) the percentage of the hits that generate SEMTs, for some circuits. Note that we inject one particle hit per μm^2 to provide more accurate and reliable analysis and confine the execution time at the same time. In Table 6, we present the distribution of the particle strikes, that is, the number of particles that affect single gates (SETs), multiple gates (SEMTs), and the number of strikes that have no impact on the circuit. This analysis is showcased for both technologies to investigate the impact of technology downscaling, indicating that the SEMTs are considerably increased at 15 nm technology and, thus, affect the evaluation process and potentially increase the SER.

Table 5. The overall number of multiple affected gates, the number of hits implemented, and the percentage of particles, which provoke SEMTs.

Benchmark	45 nm				15 nm			
	Hits	SEMTs	Affected Gates	Perc.	Hits	SEMTs	Affected Gates	Perc.
s298	100	73	259	73%	100	78	282	78%
s400	200	129	469	64%	100	81	412	81%
s5378	2300	1390	4861	60%	700	488	1124	70%
s15850	6300	4163	14,353	65%	1700	1323	7159	77%
s35932	20,000	13,075	40,619	65%	6000	5013	23,845	84%

Table 6. Distribution of SETs, SEMTs, and unaffected gates by particle strikes.

Benchmark	45 nm			15 nm		
	SETs	**SEMTs**	**Not Affected**	**SETs**	**SEMTs**	**Not Affected**
s298	16	73	11	6	78	16
s400	28	129	43	5	81	14
s5378	370	1390	540	100	488	112
s15850	1103	4163	1034	151	1323	226
s35932	4216	13,075	2709	576	5013	411

5.3. SER Estimation Results

Various simulations are performed to estimate the SER on different technologies considering the electrical and timing-masking models. Table 7 presents the SER comparison between the 45 nm and 15 nm technologies for some benchmarks, along with the corresponding average execution time. According to the technology libraries, the former is based on the conventional planar MOSFETs whereas the latter is built upon the modern non-planar FinFETs. Over the past few years, the utilization of FinFET technology in ICs fabrication has emerged as an efficient solution for potential problems due to the downscaling of device feature sizes. Among the advantages of FinFETs, this type of transistor is considered more resistant to external parameters, such as radiation. In particular, non-planar FinFET structures are not as vulnerable to heavy ions as planar transistors, resulting in smaller SET pulses induced by particle strikes, as reported in the TCAD simulations previously. However, SER probability increases for 15 nm technology as the number of SEMTs and operating frequency increase due to the smaller transistor size and the clock period reduction, respectively. On the other hand, the SER in terms of FIT decreases since its calculation incorporates the circuit area. Therefore, the probability of a particle striking a circuit designed with respect to 15 nm technology is lower as the area is smaller.

Table 7. SER evaluation of some benchmarks for the 45 nm and 15 nm technologies.

Benchmark	45 nm		15 nm		Exec. Time
	SER	**Area (μm^2)**	**SER**	**Area (μm^2)**	
s27	1.4×10^{-6}	32.31	4.7×10^{-7}	8.14	<1 s
s344	3.5×10^{-6}	213.64	1.5×10^{-6}	60.61	<1 s
s641	1.8×10^{-6}	259.26	6.5×10^{-7}	76.61	<1 s
s9234	1.8×10^{-5}	1792.54	7.5×10^{-6}	675.42	2 s
s13207	3.9×10^{-5}	6037.77	1.2×10^{-5}	1542.46	9 s
s15850	3.7×10^{-5}	6309.69	2.7×10^{-5}	1730.25	14 s
s35932	2.7×10^{-5}	19,978.45	2.1×10^{-5}	6090.48	170 s
s38584	3.1×10^{-5}	19,673.24	1.4×10^{-5}	7549.78	190 s

Table 8 demonstrates the SER probabilities on the 45 nm and 15 nm technology nodes, taking into consideration three different timing analysis approaches. The first approach to estimate gate delays is the Logical Effort (LE) technique; the second is a conventional STA based on an NLDM; and, finally, the third is an enhanced timing analysis, which incorporates an RC interconnection model to account for the parasitics delay (RC I/C). According to the experimental results, the SER probability either decreases or increases when the NLDM and RC I/C approaches are considered concerning the LE method. That is explained by the fact that LE is an approximation method to estimate gate delay, taking into account transistor widths, lengths, and the number of fanouts and inputs as well, albeit neglecting the input transition times and the actual total output load capacitance. As a result, the gate delay is overestimated or underestimated, compared to the other models, resulting in smaller or higher period values that eventually affect the evaluation of SER. At the same time, the SER accuracy for the other two approaches can be validated by the accuracy of the previously presented electrical and timing-masking models.

Table 8. SER estimation considering LE, NLDM, and RC interconnection approaches for the 45 nm and 15 nm technologies.

Benchmark	45 nm			15 nm		
	LE	NLDM	RC I/C	LE	NLDM	RC I/C
s27	0.3236	0.2348	0.2191	0.3792	0.4451	0.2832
s344	0.2089	0.1092	0.0974	0.2692	0.2937	0.1165
s641	0.0495	0.0379	0.0356	0.0514	0.0572	0.0418
s9234	0.0631	0.0568	0.0494	0.0659	0.0715	0.0548
s13207	0.0445	0.0369	0.0321	0.0511	0.0592	0.0396
s15850	0.0414	0.0342	0.0293	0.0837	0.1014	0.0772
s35932	0.0049	0.0043	0.0038	0.0127	0.0298	0.0163
s38584	0.0118	0.0061	0.0049	0.0181	0.0216	0.0116

5.4. Impact of Electrical Masking Model on SER

The comparison of the two techniques of electrical masking modeling is presented. According to the first technique, the generated pulse duration due to a particle strike is dependent on the gate delay and either attenuates or remains stable [3]. For this reason, it is considered an approximation approach, compared to the second method described in the previous section. Based on SPICE simulations, a more accurate electrical masking model is provided. More specifically, the transient glitches that may broaden through their propagation until they approach the memory elements are not taken into account by the former method, thus affecting the accuracy of SER evaluation.

SER evaluations for both methods, as well as their percentage difference, are presented in Table 9. The probability of SER is higher, using the second technique for all circuits and for both technologies, which is something reasonable since transient pulses can broaden as they propagate through a circuit according to the SPICE-oriented method. The second technique is based on STA to compute the propagation delays, making it more accurate, and faster, than the former and the time-expensive LUT-based approaches.

Table 9. SER considering an approximate pulse propagation function and a SPICE-oriented technique for the 45 nm and 15 nm technologies.

Benchmark	45 nm			15 nm		
	1st Tech.	2nd Tech.	Diff. (%)	1st Tech.	2nd Tech.	Diff. (%)
s9234	0.0327	0.0493	33%	0.0382	0.0548	34%
s13207	0.0228	0.0321	28%	0.0283	0.0396	28%
s15850	0.0172	0.0293	41%	0.0569	0.0772	26%
s35932	0.0024	0.0038	36%	0.0118	0.0163	27%
s38417	0.0324	0.0478	32%	0.0482	0.0617	21%
s38584	0.0034	0.0049	30%	0.0092	0.0116	20%

6. Conclusions

In this work, a comprehensive analysis of the electrical and timing-masking modeling is presented. The influence of these parameters on the SET pulse propagation is discussed, whereas the SET pulse generation is performed with SPICE using the current pulses obtained from TCAD characterization. Based on an integrated SER estimation tool, extensive experimental results in different technologies reveal the importance of an accurate timing analysis model to reliably evaluate modern chips. Additionally, the impact of the technology-dependent factors of SEMTs' number and operating frequency on the SER is examined. Finally, regarding the validation of the proposed models, SPICE simulations are performed indicating satisfactory accuracy.

Author Contributions: Conceptualization, P.T. and G.I.P.; data curation, P.T. and G.I.P.; funding acquisition, N.E. and G.S.; investigation, P.T. and G.I.P.; methodology, P.T. and G.I.P.; preparation, P.T. and G.I.P.; project administration, N.E. and G.S.; software, P.T. and G.I.P.; supervision, N.E. and G.S.; validation, P.T. and G.I.P.; writing—original draft, P.T. and G.I.P.; writing—review and editing, P.T., G.I.P., N.E. and G.S. All authors have read and agreed to the published version of the manuscript.

Funding: This research received no external funding.

Institutional Review Board Statement: Not applicable.

Informed Consent Statement: Not applicable.

Data Availability Statement: Not applicable.

Conflicts of Interest: The authors declare no conflict of interest.

Abbreviations

CMOS	Complementary Metal–Oxide–Semiconductor
DEF	Design Exchange Format
DTA	Dynamic Timing Analysis
EDA	Electronic Design Automation
FinFET	Fin Field-Effect Transistor
GDSII	Graphic Database System II
LE	Logical Effort
LET	Linear Energy Transfer
MOSFET	Metal–Oxide–Semiconductor Field-Effect Transistor
NLDM	Non-Linear Delay Model
PIPB	Propagation Induced Pulse Broadening
P&R	Placement and Route
SEMT	Single-Event Multiple Transient
SER	Soft Error Rate
SET	Single-Event Transient
SPEF	Standard Parasitic Exchange Format
STA	Static Timing Analysis
TCAD	Technology Computer-Aided Design
VLSI	Very-Large-Scale Integration

References

1. Shivakumar, P.; Kistler, M.; Keckler, S.W.; Burger, D.; Alvisi, L. Modeling the Effect of Technology Trends on the Soft Error Rate of Combinational Logic. In Proceedings of the International Conference on Dependable Systems and Networks (DSN'02), Washington, DC, USA, 23–26 June 2002; pp. 389–398.
2. Rossi, D.; Omana, M.; Toma, F.; Metra, C. Multiple Transient Faults in Logic: An Issue for Next Generation ICs? In Proceedings of the 20th IEEE International Symposium on Defect and Fault Tolerance in VLSI Systems (DFT'05), Monterey, CA, USA, 3–5 October 2005; pp. 352–360.

3. Paliaroutis, G.I.; Tsoumanis, P.; Evmorfopoulos, N.; Dimitriou, G.; Stamoulis, G.I. A Placement-Aware Soft Error Rate Estimation of Combinational Circuits for Multiple Transient Faults in CMOS Technology. In Proceedings of the 2018 IEEE International Symposium on Defect and Fault Tolerance in VLSI and Nanotechnology Systems (DFT'18), Chicago, IL, USA, 8–10 October 2018; pp. 1–6.

4. Cha, H.; Patel, J.H. A Logic-Level Model for α-Particle Hits in CMOS Circuits. In Proceedings of the 1993 IEEE International Conference on Computer Design (ICCD'93), Cambridge, MA, USA, 3–6 October 1993; pp. 538–542.

5. Omana, M.; Papasso, G.; Rossi, D.; Metra, C. A Model for Transient Fault Propagation in Combinatorial Logic. In Proceedings of the 9th IEEE On-Line Testing Symposium (IOLTS'03), Kos Island, Greece, 7–9 July 2003; pp. 111–115.

6. Dhillon, Y.S.; Diril, A.U.; Chatterjee, A. Soft-Error Tolerance Analysis and Optimization of Nanometer Circuits. In Proceedings of the Design, Automation and Test in Europe Conference and Exhibition (DATE), Munich, Germany, 7–11 March 2005; pp. 288–293. Volume 1.

7. Wang, F.; Xie, Y. Soft Error Rate Analysis for Combinational Logic Using an Accurate Electrical Masking Model. *IEEE Trans. Dependable Secur. Comput.* **2009**, *8*, 137–146. [CrossRef]

8. Gili, X.; Barcelo, S.; Bota, S.; Segura, J. Analytical Modeling of Single Event Transients Propagation in Combinational Logic Gates. *IEEE Trans. Nucl. Sci.* **2012**, *59*, 971–979. [CrossRef]

9. Rajaraman, R.; Kim, J.S.; Vijaykrishnan, N.; Xie, Y.; Irwin, M.J. SEAT-LA: A Soft Error Analysis Tool for Combinational Logic. In Proceedings of the 19th International Conference on VLSI Design held jointly with 5th International Conference on Embedded Systems Design (VLSID'06), Hyderabad, India, 3–7 January 2006.

10. Limbrick, D.B.; Robinson, W.H. Characterizing Single Event Transient Pulse Widths in an Open-Source Cell Library Using Spice. In Proceedings of the IEEE Workshop on Silicon Errors in Logic-System Effects (SELSE'12), Urbana-Champaign, IL, USA, 27–28 March 2012.

11. Kiamehr, S.; Ebrahimi, M.; Firouzi, F.; Tahoori, M.B. Chip-Level Modeling and Analysis of Electrical Masking of Soft Errors. In Proceedings of the IEEE 31st VLSI Test Symposium (VTS'13), Berkeley, CA, USA, 29 April–2 May 2013; pp. 1–6.

12. Watkins, A.; Tragoudas, S. An Enhanced Analytical Electrical Masking Model for Multiple Event Transients. In Proceedings of the 26th International Great Lakes Symposium on VLSI (GLSVLSI'16), Boston, MA, USA, 18–20 May 2016; pp. 369–372.

13. Wang, F.; Xie, Y. An Accurate and Efficient Model of Electrical Masking Effect for Soft Errors in Combinational Logic. In Proceedings of the 2nd Workshop on System Effects of Logic Soft Errors (SELSE'06), Urbana, IL, USA, 11–12 April 2006.

14. Dodd, P.E.; Shaneyfelt, M.R.; Felix, J.A.; Schwank, J.R. Production and Propagation of Single-Event Transients in High-Speed Digital Logic ICs. *IEEE Trans. Nucl. Sci.* **2004**, *51*, 3278–3284. [CrossRef]

15. Ferlet-Cavrois, V.; Paillet, P.; McMorrow, D.; Fel, N.; Baggio, J.; Girard, S.; Duhamel, O.; Melinger, J.S.; Gaillardin, M.; Schwank, J.R.; et al. New Insights Into Single Event Transient Propagation in Chains of Inverters—Evidence for Propagation-Induced Pulse Broadening. *IEEE Trans. Nucl. Sci.* **2007**, *54*, 2338–2346. [CrossRef]

16. Massengill, L.W.; Tuinenga, P.W. Single-Event Transient Pulse Propagation in Digital CMOS. *IEEE Trans. Nucl. Sci.* **2008**, *55*, 2861–2871. [CrossRef]

17. Wirth, G.; Kastensmidt, F.L.; Ribeiro, I. Single Event Transients in Logic Circuits—Load and Propagation Induced Pulse Broadening. *IEEE Trans. Nucl. Sci.* **2008**, *55*, 2928–2935. [CrossRef]

18. Mogollón, J.M.; Palomo, F.R.; Aguirre, M.A.; Nápoles, J.; Guzmán-Miranda, H.; García-Sánchez, E. TCAD Simulations on CMOS Propagation Induced Pulse Broadening Effect: Dependence Analysis on the Threshold Voltage. *IEEE Trans. Nucl. Sci.* **2010**, *57*, 1908–1914. [CrossRef]

19. Hamad, G.B.; Hasan, S.R.; Mohamed, O.A.; Savaria, Y. Investigating the Impact of Propagation Paths and Re-Convergent Paths on the Propagation Induced Pulse Broadening. In Proceedings of the IEEE 14th European Conference on Radiation and Its Effects on Components and Systems (RADECS'13), Oxford, UK, 23–27 September 2013; pp. 1–4.

20. Chi, Y.; Song, R.; Shi, S.; Liu, B.; Cai, L.; Hu, C.; Guo, G. Characterization of Single-Event Transient Pulse Broadening Effect in 65 nm Bulk Inverter Chains Using Heavy Ion Microbeam. *IEEE Trans. Nucl. Sci.* **2016**, *64*, 119–124. [CrossRef]

21. Black, D.A.; Robinson, W.H.; Wilcox, I.Z.; Limbrick, D.B.; Black, J.D. Modeling of Single Event Transients With Dual Double-Exponential Current Sources: Implications for Logic Cell Characterization. *IEEE Trans. Nucl. Sci.* **2015**, *62*, 1540–1549. [CrossRef]

22. Asadi, H.; Tahoori, M.B. Soft Error Derating Computation in Sequential Circuits. In Proceedings of the 2006 IEEE/ACM International Conference on Computer Aided Design (ICCAD), San Jose, CA, USA, 5–9 November 2006; pp. 497–501.

23. Tsoumanis, P.; Paliaroutis, G.I.; Evmorfopoulos, N.; Stamoulis, G. On the Impact of Electrical Masking and Timing Analysis on Soft Error Rate Estimation in Deep Submicron Technologies. In Proceedings of the 2021 IEEE International Symposium on Defect and Fault Tolerance in VLSI and Nanotechnology Systems (DFT'21), Athens, Greece, 6–8 October 2021; pp. 1–6.

24. Elmore, W.C. The Transient Response of Damped Linear Networks with Particular Regard to Wideband Amplifiers. *Int. J. Appl. Phys.* **1948**, *19*, 55–63. [CrossRef]

25. Open-Cell Library—Silicon Integration Initiative. Available online: https://si2.org/open-cell-library/ (accessed on 20 January 2022).

26. Paliaroutis, G.I.; Tsoumanis, P.; Evmorfopoulos, N.; Dimitriou, G.; Stamoulis, G. SET Pulse Characterization and SER Estimation in Combinational Logic with Placement and Multiple Transient Faults Considerations. *Technologies* **2020**, *8*, 5. [CrossRef]

 *technologies*

Article

Efficient Stochastic Computing FIR Filtering Using Sigma-Delta Modulated Signals

Nikos Temenos *, Anastasis Vlachos and Paul P. Sotiriadis

Department of Electrical and Computer Engineering, National Technical University of Athens,
157 80 Athens, Greece; vlahosanastasis@gmail.com (A.V.); pps@ieee.org (P.P.S.)
* Correspondence: ntemenos@gmail.com

Abstract: This work presents a soft-filtering digital signal processing architecture based on sigma-delta modulators and stochastic computing. A sigma-delta modulator converts the input high-resolution signal to a single-bit stream enabling filtering structures to be realized using stochastic computing's negligible-area multipliers. Simulation in the spectral domain demonstrates the filter's proper operation and its roll-off behavior, as well as the signal-to-noise ratio improvement using the sigma-delta modulator, compared to typical stochastic computing filter realizations. The proposed architecture's hardware advantages are showcased with synthesis results for two FIR filters using FPGA and synopsys tools, while comparisons with standard stochastic computing-based hardware realizations, as well as with conventional binary ones, demonstrate its efficacy.

Keywords: stochastic computing; stochastic filtering; stochastic FIR filter; unconventional computing

Citation: Temenos, N.; Vlachos, A.; Sotiriadis, P.P. Efficient Stochastic Computing FIR Filtering Using Sigma-Delta Modulated Signals. *Technologies* **2022**, *10*, 14. https://doi.org/10.3390/technologies10010014

Academic Editors: Manoj Gupta, Spiros Nikolaidis and Rodrigo Picos

Received: 4 November 2021
Accepted: 10 January 2022
Published: 20 January 2022

Publisher's Note: MDPI stays neutral with regard to jurisdictional claims in published maps and institutional affiliations.

1. Introduction

Modern digital signal processing (DSP) blocks are characterized by hardware efficiency and high-performance computations [1,2]. On the other hand, standard binary computing methods impose constraints on their design specifications, namely power, area, and energy, which are continuously increasing given the rise of hardware-taxing emerging applications [2]. To this end, unconventional computing paradigms are explored as an alternative to the binary one, with stochastic computing (SC) being an attractive approach [3–6].

SC represents real-valued numbers in the form of stochastic sequences [7]. Encoding the information in such a way makes its processing resilient to soft errors originating from noisy sources [3,8], for instance bit-flips, which is prohibitive in the case of the binary arithmetic. Furthermore, SC's bit-processing nature allows for the realization of fundamental arithmetic operations as well as highly-complex functions using a few standard logic gates and cells, thereby reducing the hardware requirements when compared to their binary counterparts [3,4,9–12].

The advantages of SC favor the design of several applications, especially the ones in which parallelization is necessary. These include neural networks [13–19], digital image processing [20–24], soft-polynomial solving and filtering [25–30], modern error coding and decoding, etc. [31]. Focusing on digital filters, SC-based implementations of the standard Nyquist-rate ones are proven to be hardware-efficient [25,27–30]. Their key advantage lies in the multiplication process; instead of using the area-demanding binary multipliers, SC exploits simple XNOR gates and MUXs to realize the multiplication and addition parts, respectively, improving the area occupied by the intensive multiply-and-add operations.

Although the SC filters realized using the methods considered in [27–30] reduce the hardware resources when compared to the standard binary approach, they are also limited in their spectral characteristics. Specifically, their performance is affected by two factors: (1) the length of the stochastic sequence required to process a single sample, which is SC's essential design trade-off, and (2) the noise introduced from the binary-to-stochastic converters used to generate the input signal and the filter's coefficients.

A well-known method to reduce the noise floor and improve the spectral characteristics of a quantized signal is to use sigma-delta modulators (SDMs) [32–34]. They convert a high-resolution signal (several bits) into a lower-bit one by employing the technique of oversampling; the input signal is sampled at a frequency much higher than the Nyquist, thus reducing the noise in the desired frequency band of interest.

Motivated by the properties of SDMs, the use of a first-order SDM in FIR filtering was recently explored in [26]. It serves as a single-bit encoder of an input (quantized) signal, allowing for the SC-based FIR filter's multipliers to be realized exclusively by simple logic gates. As such, the proposed SDM-SC architecture's advantage is twofold; on one hand it offers improved signal-to-noise ratio (SNR), due to the SDM's oversampling technique, which is not possible with conventional SC-based filtering; on the other hand, the SDM allows for the filtered signal to be *encoded in time*, therefore reducing the SC's typical accuracy–latency trade-off as well as the power and energy consumed for this process.

We organize the remainder of the proposed work as follows. In Section 2, we provide a background on the stochastic numbers and their properties, as well as the operation of the first-order SDM. In Section 3, we review the prior work in SC-based FIR filter realizations and mathematically formulate their operation. In Section 4, we present the SDM-SC architecture and explain its principle operation through proper analysis. Section 5 includes experimental results with respect to (1) the SDM-SC's architecture spectral characteristics and (2) comparisons with the SC literature as well as the conventional binary filters in SNR, FPGA synthesis results, and hardware resources in a 45 nm process. Finally, Section 6 provides the conclusion.

2. Stochastic Computing and Sigma-Delta Modulation Notation and Principle Operation

In this section, we provide a background on the generation of stochastic numbers and their manipulation using standard logic gates, as well as explain the operation of the first-order sigma-delta modulator.

2.1. Stochastic Number Generation & Properties

The conversion of a binary number into a stochastic one is typically performed by the stochastic number generator (SNG), shown in Figure 1. Its operation is based upon the sampling on each clock cycle of a pseudo-random number generator uniformly distributed in $\mathcal{R}_s = \{0, 1, \ldots, 2^k - 1\}$, with the desired binary number $B \in [0, 1]$ of the same bit length. Usually, the pseudo-random number source is implemented as a k-bit linear feedback shift register (LFSR), but note that other methods can also be used [35]. The bit generation is completed after $N = 2^k$ clock cycles and corresponds to the length of the stochastic sequence.

Formally, the generated sequence of length N, $\{X_n\}_{n=1}^N$, where n denotes the current sample processed (or the current clock cycle), is assumed to be an independent and identically distributed (IID) Bernoulli sequence. Therefore, the generated stochastic number (SN) has probability defined as

$$X \triangleq P(X_n = 1) \tag{1}$$

and mean value

$$\tilde{X}_N \triangleq \frac{1}{N} \sum_{n=1}^N X_n. \tag{2}$$

The SN's mean value $\tilde{X}_N$ represents a non-negative number in $[0, 1]$, known as *unipolar format* in SC, whereas to obtain a negative SN representation (known as *bipolar format*), the transformation $X \mapsto 2X - 1$ is used, expanding the range of the SN to the interval $[-1, 1]$. As expected, in both formats the sequence length N plays a critical role in the accuracy of the SN given the fact that it increases at the cost of additional clock cycles and is inversely proportional to $\sqrt{N}$ [36]. To further investigate the equivalent noise introduced by an SN, one can consider the noise figure (NF) [37] defined as

$$NF \triangleq 10 \log 10 \left(\frac{P_S}{P_N} \right),$$ (3)

where P_S and P_N are the average power and noise, respectively, of the generated SN.

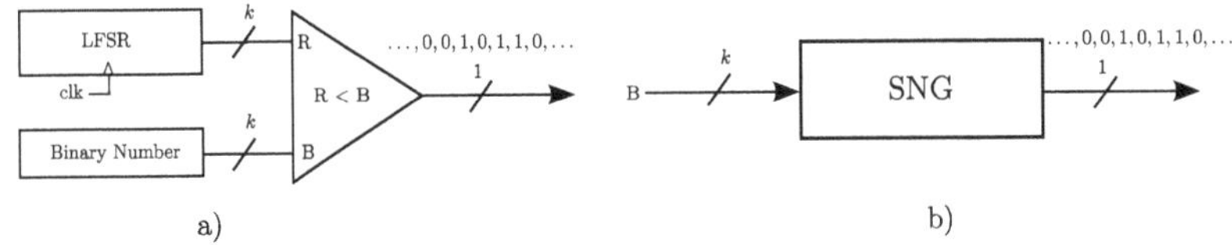

Figure 1. Stochastic sequence generation: (**a**) Stochastic number generator (SNG) block; (**b**) Equivalent block.

2.2. Mathematical Properties of Logic Gates in Stochastic Computing

Fundamental mathematical operations are supported in the context of SC and can be implemented by simple logical gates, according to the format used. For the following, we assume that $\{X_n\}_{n=1}^N$, $\{Y_n\}_{n=1}^N$ are stochastic sequences generated by different SNGs, to ensure independence among them, and $\{H_n\}_{n=1}^N$ is the result of their operation. We also note that whenever the bipolar format is required, the transformation $X \mapsto 2X - 1$ is used.

- NOT Gate
  In unipolar format, the output of the NOT gate, $H_n = \text{NOT}(X_n)$, complements the probability of the input,

$$H = P(H_n = 1) = P(X_n = 0) = 1 - P(X_n = 1) = 1 - X,$$ (4)

  whereas in the bipolar format, it operates as a sign inverter.

$$H = P(H_n = 1) = P(X_n = 0) = 1 - P(X_n = 1) = -X.$$ (5)

- AND Gate:
  The AND gate in unipolar format, $H_n = \text{AND}(X_n, Y_n)$, performs multiplication.

$$H = P(H_n = 1) = P(X_n = 1, Y_n = 1) = P(X_n = 1)P(Y_n = 1) = XY.$$ (6)

- XNOR Gate:
  The XNOR gate in bipolar format, $H_n = \text{XNOR}(X_n, Y_n)$, performs multiplication.

$$\begin{aligned} H = P(H_n = 1) &= P(X_n = 1, Y_n = 1) + P(X_n = 0, Y_n = 0) \\ &= 2P(X_n = 1)P(Y_n = 1) - P(X_n = 1) - P(Y_n = 1) + 1 \\ &= XY. \end{aligned}$$ (7)

- Multiplexer
  Assuming an an IID control sequence $\{C_n\}_{n=1}^N$, the multiplexer (MUX), $H_n = \text{MUX}(X_n, Y_n; C_n)$, is the standard way to perform scaled addition between two SN, regardless of the format used, and is given as

$$\begin{aligned} H = P(H_n = 1) &= P(X_n = 1, C_n = 1) + P(Y_n = 1, C_n = 0) \\ &= P(X_n = 1)P(C_n = 1) + P(Y_n = 1)P(C_n = 0) \\ &= XC + Y\overline{C}. \end{aligned}$$ (8)

Furthermore, if $P(C_n = 1) = 1/2$, the MUX operates as a scaling adder, i.e.,

$$H = P(H_n = 1) = \frac{P(X_n = 1) + P(Y_n = 1)}{2} = \frac{X + Y}{2}. \tag{9}$$

Stochastic subtraction, on the other hand, can only be realized in the bipolar format, using a NOT gate in one of the two inputs as

$$H = P(H_n = 1) = \frac{P(X_n = 1) + P(Y_n = 0)}{2} = \frac{X - Y}{2}. \tag{10}$$

The operation of logic gates with specific interest is illustrated with an example in Figure 2.

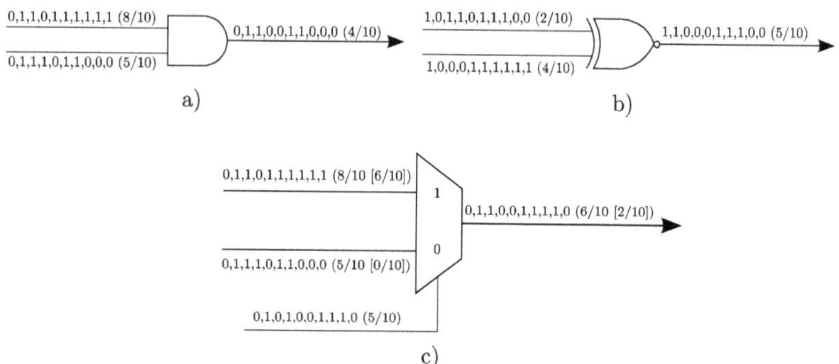

Figure 2. Stochastic computing fundamental operations: (**a**) Unipolar multiplication, (**b**) bipolar multiplication, (**c**) unipolar/bipolar addition.

2.3. Correlation in Stochastic Computing

The proper operation of SC processing elements is based upon the assumption that different SNGs are used as their input sequences. This is due to the fact that using the same initial seed in their LFSRs will cycle through the values in $\mathcal{R}_S$ simultaneously, thereby generating maximally correlated sequences (overlap of logic 1s) given the same binary number for conversion [38–40]. To provide better insight on the former, consider the case where the multiplication of two SNs is desired, i.e., $H = XY$, with $X = Y = 0.6$. If SNGs with the same initial LFSR seed are used, then two identical sequences will be generated and their multiplication will result in $H = 0.6$, instead of the correct one $H = 0.36$. The same also happens when the LFSR is shared among the SNGs, without different seed initialization or proper use of delays. However, it has been shown that in certain cases, maximally correlated sequences can benefit specific applications [3,10], offering promising results.

A standard measure of correlation in SC is the stochastic computing correlation (SCC) [3,38]. For any two sequences $\{X_n\}$ and $\{Y_n\}$, then $\mathrm{SCC}(X_n, Y_n)$ is calculated as

$$\mathrm{SCC}(X_n, Y_n) = \begin{cases} \dfrac{\mathbb{E}[X_n, Y_n] - \mathbb{E}[X_n]\mathbb{E}[Y_n]}{\min(\mathbb{E}[X_n], \mathbb{E}[Y_n]) - \mathbb{E}[X_n]\mathbb{E}[Y_n]}, & \mathbb{E}[X_n, Y_n] > \mathbb{E}[X_n]\mathbb{E}[Y_n] \\[2em] \dfrac{\mathbb{E}[X_n, Y_n] - \mathbb{E}[X_n]\mathbb{E}[Y_n]}{\mathbb{E}[X_n]\mathbb{E}[Y_n] - \max(\mathbb{E}[X_n] + \mathbb{E}[Y_n] - 1, 0)}, & \text{otherwise} \end{cases}, \tag{11}$$

taking values in $[-1, 1]$, with $\mathrm{SCC}(X_n, Y_n) = 0$ corresponding to uncorrelated sequences. Note that SCC can also be used to measure the the auto-correlation of an output sequence, assuming the current output sample H_n and its delayed version by $r > 0$ samples H_k, with $k = n + r$, as $\mathrm{SCC}(H_n, H_k)$.

2.4. The First-Order Sigma-Delta Modulator

A Sigma-Delta Modulator (SDM) is typically used to convert a higher-resolution analog or digital signal into a lower-bit one. Its main advantage is the exploitation of the oversampling technique, which pushes the in-band quantization noise outside the input signal's frequency band of interest. This is accomplished by sampling the input signal at a rate f_s much higher than the Nyquist one. The oversampling ratio (OSR) is defined as

$$OSR \triangleq \frac{f_s}{2f_B}, \tag{12}$$

where f_B is the maximum input signal's frequency. Oversampling has a direct impact on the spectral quality of the modulator and, more specifically, increasing the OSR leads to an improvement of the modulator's signal-to-noise ratio (SNR) [34].

The first-order single-bit SDM is shown in Figure 3. It comprises an adder and an integrator, followed by a two-step quantizer block, $Q(\cdot)$, which behaves as a nonlinear function as

$$Q(Y_n) = \begin{cases} +1, Y_n \geq 0 \\ -1, Y_n < 0 \end{cases}. \tag{13}$$

According to the SDM of Figure 3, the modulator's input U_n, with n being the time index, is of m-bits length, whereas its output V_n is single-bit ± 1. Therefore, the behavior of the first-order SDM can be expressed as

$$V_n = Q(Y_{n-1} + U_n - V_{n-1}). \tag{14}$$

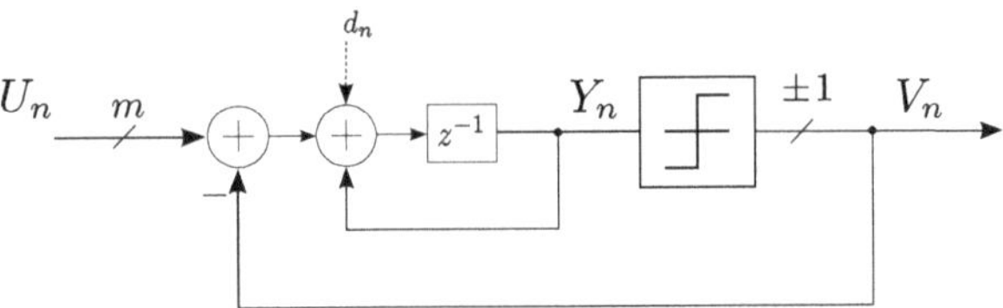

Figure 3. The first-order single-bit sigma-delta modulator. A dithering sequence can optionally be used.

3. Prior Work in Stochastic Computing FIR Filters

For an arbitrary discrete-time signal U_t, $t = 1, 2, \ldots, T$ and filter coefficients w_m, $m = 1, 2 \ldots, M$, an M-tap finite impulse response (FIR) filter is described as

$$Z_t = \sum_{m=0}^{M-1} w_m V_{t-m} = \boldsymbol{w}^T \boldsymbol{V}. \tag{15}$$

where $\boldsymbol{w} = [w_0, w_1, \ldots, w_{M-1}] \in \mathbb{R}^M$ and $\boldsymbol{V} = [V_t, V_{t-1}, \ldots, V_{t-M+1}] \in \mathbb{R}^M$. A straightforward realization of Equation (15), using SC techniques, requires $M - 1$ XNOR gates for multiplication, as the values of w_m, and V_{t-m} might be negative, and an M-to-1 MUX for addition. However, this implementation leads to the downscaling of the output by a factor of $1/M$, which causes severe accuracy loss, especially when the filter's order M is large.

To address this problem, a stochastic FIR filter that uses an MUX adder-tree was proposed in [41], and is based on the inner-product processing unit shown in Figure 4. Instead of the standard method to perform multiplication in bipolar format using XNOR gates, the sign of the weights is also considered, and thus the multiplications are realized using XOR gates. To explain the signed XOR gate's operation as an SC multiplier in bipolar format, assume first a 2s complement binary representation of a signed-value weight w_m, where its most significant bit serves as its sign, i.e.,

$$sign(w_m) = \begin{cases} 0, w_m \geq 0 \\ 1, w_m < 0 \end{cases}. \tag{16}$$

Considering now a sample of the input signal V_t converted into a stochastic sequence with probability $P(V_{t,n} = 1)$, then the output of an XOR gate is

$$P(G_{t,n} = 1) = sign(w_m) + P(V_{t,n} = 1) - 2sign(w_m)P(V_{t,n} = 1)$$
$$= \begin{cases} 1 - P(V_{t,n} = 1), & sign(w_m) = 1 \\ P(V_{t,n} = 1), & sign(w_m) = 0' \end{cases} \tag{17}$$

which is simplified to a multiplication $sign(w_m)V_t$ given the definitions of the NOT gate in Section 2. Finally, using an uneven control signal with probability $P(C_n = 1) = |w_0|/(|w_0| + |w_1|)$, the output of the MUX is

$$Z = P(Z_n = 1) = P(C_n = 1)P(G_{t,n} = 1) + P(C_n = 0)P(G_{t-1,n} = 1)$$
$$= \left(\frac{|w_0|}{|w_0| + |w_1|}\right)sign(w_0)V_t + \left(1 - \frac{|w_0|}{|w_0| + |w_1|}\right)sign(w_1)V_{t-1} \tag{18}$$
$$= \frac{1}{|w_0| + |w_1|}\left(w_0 V_t + w_1 V_{t-1}\right),$$

producing an inner-product scaled by $1/(|w_0| + |w_1|)$.

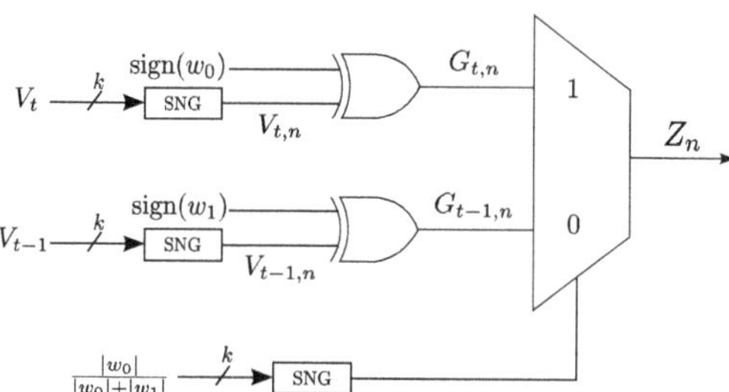

Figure 4. Inner-product block with two inputs.

Based on the inner product module of Figure 4 and the former analysis, an M-tap stochastic FIR filter can be realized, with its output scaled, however, by a factor of $1/\sum_{m=0}^{M-1}|w_m|$. It requires, in total, M XNOR gates, $M - 1$ MUXs, and $2M - 1$ SNGs. A representative example of a five-tap FIR filter implemented with the aforementioned inner-product module is shown in Figure 5.

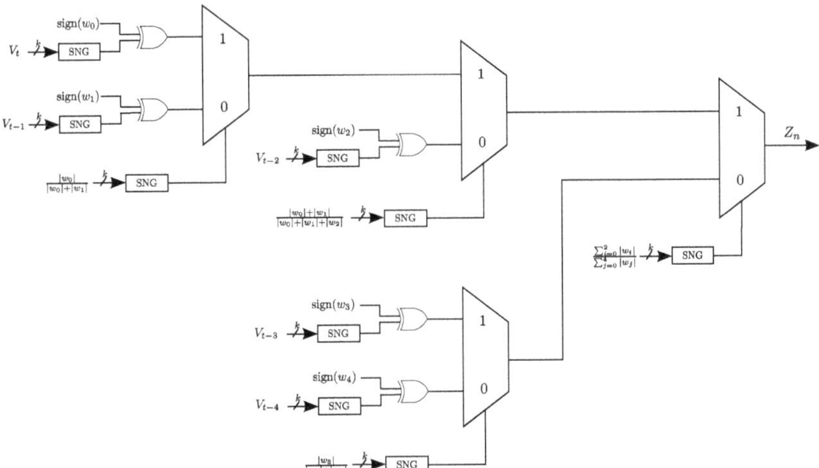

Figure 5. Stochastic implementation of a five-tap FIR filter based on the inner-product block of Figure 4.

4. Proposed SDM-SC Processing Scheme

In this section, we present the proposed SDM-SC architecture as it was introduced in [26], and is shown in Figure 6. It converts a multi-bit input signal into a single-bit one using a first-order SDM to exploit the SC's encoding and benefit from its low-area advantages. Moreover, it allows for *time-encoding* and, consequently, processing of the input signal, therefore bypassing the long-latency of the standard SC approaches. To proceed with the detailed analysis of the architecture, we start from the first-order SDM.

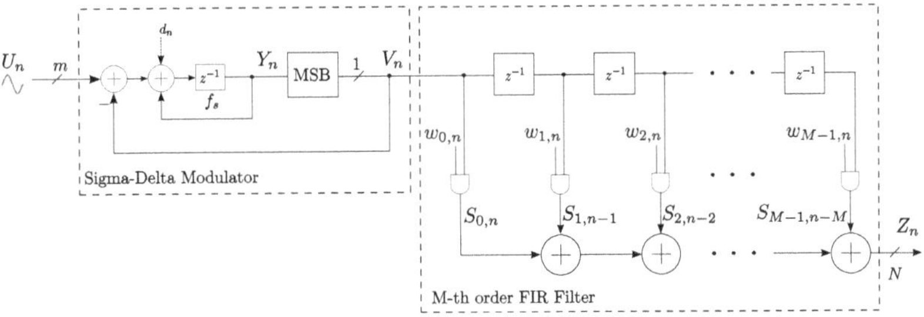

Figure 6. Proposed SDM-SC architecture. The first-order SDM encodes a multi-bit input signal into a single-bit one, carrying the information in 0, 1 representation. The sequence is then processed by an SC-based M-tap FIR filter. A dithering sequence can optionally be used.

4.1. SDM Encoding

In the architecture of Figure 6, the SDM block is the digital realization of the system-level one shown in Figure 3. Its input U_n is of m-bits length, whereas V_n is the single-bit output. A register of c-bits can replace the integrator of Figure 3, allowing for the SDM's iterative behavior to be expressed according to Equation (14). With respect to its size (in bits), it should be noted that $c \geq m$ so as to account for the accumulation process, with typical value $c = m + 1$.

The quantization process of the SDM in Figure 3 can be modeled simply as the register's most significant bit (MSB); the current input sample, i.e., U_n, determines if, and only if, the accumulator's current value is positive or negative, corresponding to an MSB of 0 or 1, respectively. Finally, it should be noted that V_n's negative value, i.e., 1, is converted into -1 using sign-extension methods.

The SDM's maximum operating frequency f_s corresponds to the register's one and, as expected, determines the input signal's maximum operating frequency f_B that the architecture is able to process. Optionally, a dithering sequence can be employed to further decrease the SDM's output noise floor [32–34].

4.2. Stochastic FIR Filter

According to Equation (15), the binary implementation of an M-tap FIR filter requires $M - 1$ D flip-flops, M binary multipliers of $m + l$-bit length, where m and l are the input signal's and the coefficient's bit resolutions, respectively, and a binary adder of $m + l + log_2(M) - 1$ bits to avoid overflows, based on the guidelines in [42].

The architecture of Figure 6 exploits the SDM's $0, 1$ encoding of the input signal, allowing for the M binary multipliers of $m + l$-bit length to be replaced by M AND gates. Furthermore, in contrast to the standard adder method used in SC, which is based upon the realization of an adder tree using the inner-product processing block of Figure 4, we use a simple binary adder of $N = \lceil log_2 M \rceil$-bits. As such, the value of Z_n is binary and belongs in $\{0, 1, \ldots, M - 1\}$. At this point, we note that one can also explore single-bit output implementations of the proposed architecture, using several non-scaling adders available in the SC literature [11,20,43].

To proceed with further analysis, we assume that each weight is converted into a stochastic sequence with probability $w_m = P(w_{m,n} = 1)$, with $m = 1, 2, \ldots, M$, and also that the probability of each AND gate's output is

$$S_m = P(S_{m,n} = 1) = P(w_{m,n} = 1)V_n. \tag{19}$$

Considering the above and the architecture of Figure 6, the instantaneous value of the output Z_n is the sum of the multiplications and is given as

$$Z_n = \sum_{m=0}^{M-1} S_m. \tag{20}$$

4.3. Stochastic Coefficient Generation

The conversion of the filter's coefficients w_m to stochastic numbers in the architecture of Figure 6 requires M SNGs. This, however, implies the use of M LFSRs of k-bits, which, along with their respective k-bit comparators, are hardware-taxing (in total, $k \times M$ registers). On the other hand, simple sharing of a single LFSR, as the random number generator between all SNGs, introduces maximal correlation among the generated sequences according to Equation (11), and thus it is expected to degrade the filtered signal's spectral quality [26,41].

In order to reduce the hardware resources from the SNGs, we employ the LFSR circular shifting scheme proposed in [44], shown in Figure 7. This scheme generates M stochastic sequences in parallel without being maximally correlated using a single LFSR. This is achieved, since the LFSR cycles through all of its values within $\mathcal{R}_s$ only once. Therefore, if $R_{n,i}$ is the LFSR's current binary value n at time index $i = 1, 2, \ldots, M$, the circular shift by $s \in \mathbb{N}^*$ bits, where $s < k$, produces the next value $R_{n,i+1}$ as

$$R_{n,i+1} \triangleq R_{(n-s,i)_N} = R_{n-s,i} \mod N. \tag{21}$$

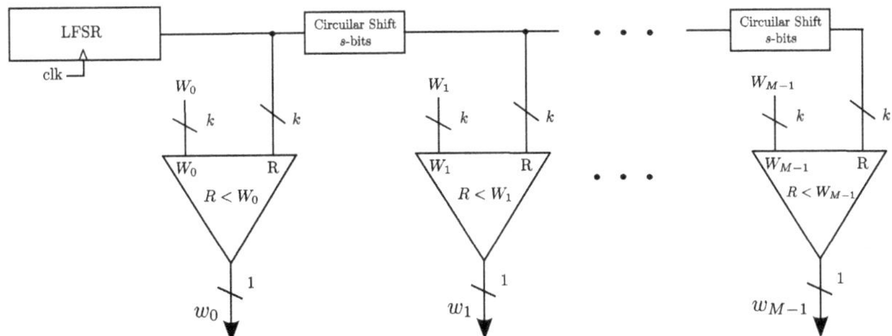

Figure 7. SNG sharing scheme with circular shifting.

5. Experimental Results

In this section, we demonstrate the performance of the proposed SDM-SC architecture. Specifically, we show experimental results with respect to 1) its spectral characteristics, and 2) comparisons with standard SC-based approaches, as well as with the conventional binary, in SNR and hardware resources.

5.1. Performance of the Proposed SDM-SC Architecture in the Spectral Domain

Here, we evaluate the performance of the proposed SDM-SC architecture in the spectral domain with simulations using MATLAB. We consider a sinusoidal input signal $U_n = sin(2\pi f_B n)$, and test it over two FIR filters, a 5- and a 7-tap one. The simulation parameters, including the filters' weights, are summarized in Table 1.

It is important to note that all frequencies are fractional with respect to the sampling frequency f_s. Moreover, to generate the sequences of the weights, LFSRs with initial register size of 15-bits are used and correspond to sequences with length 2^{15}.

Table 1. Simulation parameters.

Parameter Name	Parameter Value
Input signal U_n	$sin(2\pi f_B n)$
5-tap FIR filter weights	$w_0 = w_4 = 0.7,\ w_1 = w_3 = 0.6,\ w_2 = 0.9$
7-tap FIR filter weights	$w_0 = w_6 = 0.6,\ w_1 = w_5 = 0.4,\ w_2 = w_4 = 0.3,\ w_3 = 0.9$

In Figures 8 and 9, the power spectral density (PSD) (top) and frequency response (bottom) for the 5 and 7-tap FIR filters are respectively shown. To calculate the power spectral density, MATLAB's pwelch function was used and 10^6 samples were considered. With respect to the frequency response, it can be observed that the SDM-SC architecture follows the conventional one's, correctly achieving the cut-off frequency $\omega_{-3\text{dB}}$.

To showcase the ability of the proposed SDM-SC architecture to achieve improved performance, we show the power spectral density of the SDM's output in Figure 10. It is calculated using MATLAB's pwelch function with 10^6 samples. As one can observe the SDM reduces the noise floor in low frequencies.

additionally 669 LUTs and 25 slice registers (in total, 698 LUTs and 60 slice registers), making the SDM-SC architecture a hardware-favored approach as it reduces the LUTs by 96% and the slice registers by 42% of the binary filter.

To realize the 7-tap filter, the SDM-SC architecture requires only two registers and one LUT more; increasing the order of the filter by two requires two delays, corresponding to two flip-flops as a result of the SDM's single bit encoding, while the one LUT increase is due to the additional wiring required to output the result. The inner-product approach, however, requires 60 slice registers more to increase the filter's order by two, which is due to the four SNGs required for the addition of an inner-product block as well as the MUX producing the output. The increase on the hardware resources is also observed when the conventional binary FIR filter's order is increased, which is equal to 209 LUT and 30 register slices more. As such, the SDM-SC architecture reduces, in this case, the binary realization's LUT and slice register utilization by 96.7% and 60%, respectively.

5.4. Hardware Resources Comparison in a 45 nm Technology

To further proceed with the hardware comparisons, here we provide the resources required to realize the two FIR filters with 5 and 7 taps, extracted using the Synopsys Design Compiler with the FreePDK CMOS library at 45 nm [45]. For the comparisons, the following estimates are provided: (1) the total area in (μm^2); (2) the average power consumption in mW; (3) the delay in ns; (4) the energy consumption in pJ, defined as the average power $\times$ delay product. In Table 4, the results are shown in detail.

Table 4. Comparison of hardware resources for the realization of two FIR filters with 5 & 7 taps in area (μm^2), power (mW), delay (ns), and energy (pJ).

	5-Tap FIR Filter		
	SDM-SC Filter	**Inner-Product Adder-Tree [29,30,41]**	**Conv. Binary**
Area (μm^2)	617.12	3425.88	9853.89
Power (mW)	0.75	2.97	5.19
Delay (ns)		1.5	
Energy (pJ)	1.13	4.46	7.78
	7-tap FIR filter		
	SDM-SC Filter	Inner-Product Adder-Tree [29,30,41]	Conv. Binary
Area (μm^2)	700.66	4935.67	13,428.55
Power (mW)	0.83	4.39	7.12
Delay (ns)		1.5	
Energy (pJ)	1.24	6.51	10.58

Focusing on the realization of the 5-tap FIR filter, when compared to the inner-product and conventional binary approaches, the proposed SDM-SC architecture reduces, respectively, 1) the total area occupied by 81.9% and 93.7%, and 2) the total power and energy consumption by 74.7% and 85.5%. For the realization of the 7-tap FIR filter, the proposed SDM-SC architecture reduces, respectively, 1) the total area occupied by 85.8% and 94.8%, and 2) the total power and energy consumption by 81.09% and 88.34%. As such, from Table 4, one can conclude that the proposed SDM-SC architecture is the most efficient, hardware-wise, as it results in the least occupied area and power and energy consumption among the considered approaches.

Of great interest is the number of resources required to increase the filter's order. The SDM-SC architecture requires only 83.54 μm^2, 0.08 mW, and 0.11 pJ, corresponding to an increase of the resources by 11.9%, 9.6%, and 8.8%, respectively. On the other hand, the inner-product approach requires 1509.8 μm^2, 1.42 mW, and 2.05 pJ, corresponding to an increase of the resources by 30.5%, 32.3%, and 31.4%, respectively, whereas the conventional binary requires 3574.7 μm^2, 1.93 mW, and 2.8 pJ, corresponding to an increase of the resources by 26.6%, 27.1%, and 26.4%, respectively.

A further advantage of the SDM-SC's architecture in filtering over the inner-product approach is that of time-encoding. Assuming that the input signal U_n is of length $\hat{N}$, then the SDM-SC approach processes $\hat{N}$ samples in N clock cycles, whereas the inner-product approach requires N clock cycles for a single sample of $\hat{N}$, thus requiring $N \times \hat{N}$ clock cycles to complete the processing. This, however, reflects on the total dissipated energy by the inner-product approach, which is increased by N. Therefore, this makes the SDM-SC architecture both faster in terms of processing time and a more energy-efficient approximate processing approach, besides its improved SNR performance.

6. Conclusions

A soft-filtering architecture based on SDMs and SC was presented in this work. The first-order SDM acts as a single-bit encoder to benefit from the negligible area SC multipliers, implemented as XNOR gates. The performance of the SDM-SC architecture was evaluated in two different-order FIR filters in the spectral domain, where the filters' proper operation and roll-off behavior were shown. Compared to the standard SC-based filtering approach, it was shown that the SDM-SC architecture improves the SNR by more than 10 dB, while at the same time eliminates the SC's typical latency–accuracy trade-off, as it provides encoding in time, reflecting on the power and energy consumption, as well. With respect to the hardware resources, FPGA and synopsys synthesis results demonstrated the SDM-SC's negligible area advantages and its highly energy-efficient processing over standard SC-based and binary approaches.

Author Contributions: Conceptualization, N.T. and P.P.S.; methodology, N.T. and A.V.; software, A.V. and N.T.; validation, N.T.; writing—original draft preparation, N.T.; writing—review and editing, N.T., A.V., and P.P.S.; visualization, N.T. and A.V.; supervision, P.P.S. All authors have read and agreed to the published version of the manuscript.

Funding: The research work was supported by the Hellenic Foundation for Research and Innovation (HFRI) under the HFRI PhD Fellowship grant (Fellowship Number:1216).

Institutional Review Board Statement: Not applicable.

Informed Consent Statement: Not applicable.

Data Availability Statement: Not applicable.

Acknowledgments: The authors would like to thank anonymous reviewers for their kind suggestions and comments.

Conflicts of Interest: The authors declare no conflicts of interest.

Abbreviations

The following abbreviations are used in this manuscript:

FIR	Finite impulse response
FPGA	Field-programmable gate-array
LFSR	Linear-feedback shift register
LUT	Look-up table
MUX	Multiplexer
OSR	Oversampling ratio
PSD	Power Spectral Density
SC	Stochastic computing
SDM	Sigma-delta modulator
SDM-SC	Sigma-delta modulator-stochastic computing
SN	Stochastic number

SNG	Stochastic number generator
SNR	Signal-to-noise ratio
XNOR	Exclusive-NOR
XOR	Exclusive-OR

References

1. Mounica, Y.; Kumar, K.N.; Veeramachaneni, S.; Mahammad, N. Energy efficient signed and unsigned radix 16 booth multiplier design. *Comput. Electr. Eng.* **2021**, *90*, 106892. [CrossRef]
2. Leon, V.; Paparouni, T.; Petrongonas, E.; Soudris, D.; Pekmestzi, K. Improving Power of DSP and CNN Hardware Accelerators Using Approximate Floating-point Multipliers. *ACM Trans. Embed. Comput. Syst.* **2021**, *20*, 1–21. [CrossRef]
3. Alaghi, A.; Qian, W.; Hayes, J.P. The Promise and Challenge of Stochastic Computing. *IEEE Trans. Comput.-Aided Des. Integr. Circuits Syst.* **2018**, *37*, 1515–1531. [CrossRef]
4. Gross, W.J.; Gaudet, V.C. *Stochastic Computing: Techniques and Applications*; Springer, International Publishing, Springer Nature Switzerland AG: Cham, Switzerland, 2019.
5. Alaghi, A.; Hayes, J.P. Survey of Stochastic Computing. *ACM Trans. Embed. Comput. Syst.* **2013**, *12*, 1–19. [CrossRef]
6. Metku, P.; Seva, R.; Choi, M. Energy-Performance Scalability Analysis of a Novel Quasi-Stochastic Computing Approach. *MDPI J. Low Power Electron. Appl.* **2019**, *9*, 30. [CrossRef]
7. Gaines, B.R. *Stochastic Computing Systems*; Springer: Boston, MA, USA, 1967.
8. Najafi, M.H.; Jenson, D.; Lilja, D.J.; Riedel, M.D. Performing Stochastic Computation Deterministically. *IEEE Trans. Very Large Scale Integr. (VLSI) Syst.* **2019**, *27*, 2925–2938. [CrossRef]
9. Temenos, N.; Sotiriadis, P.P. Deterministic Finite State Machines for Stochastic Division in Unipolar Format. In Proceedings of the IEEE International Symposium on Circuits and Systems (ISCAS), Seville, Spain, 12–14 October 2020.
10. Lee, V.T.; Alaghi, A.; Hayes, J.P.; Sathe, V.; Ceze, L. Energy-efficient hybrid stochastic-binary neural networks for near-sensor computing. In ACM Proceedings of the Conference on Design, Automation & Test in Europe, Laussane, Switzerland, 27–31 March 2017.
11. Yuan, B.; Wang, Y.; Wang, Z. Area-Efficient Scaling-Free DFT/FFT Design Using Stochastic Computing. *IEEE Trans. Circuits Syst. II Express Briefs* **2016**, *63*, 1131–1135. [CrossRef]
12. Ting, P.; Hayes, J.P. Eliminating a hidden error source in stochastic circuits. In Proceedings of the IEEE International Symposium on Defect and Fault Tolerance in VLSI and Nanotechnology Systems (DFT), Cambridge, UK, 23–25 October 2017.
13. Morro, A.; Canals, V.; Oliver, A.; Alomar, M.L.; Galán-Prado, F.; Ballester, P.J.; Rosselló, J.L. A Stochastic Spiking Neural Network for Virtual Screening. *IEEE Trans. Neural Netw. Learn. Syst.* **2018**, *29*, 1371–1375. [CrossRef] [PubMed]
14. Brown, B.D.; Card, H.C. Stochastic Neural Computation I: Computational Elements. *IEEE Trans. Comput.* **2002**, *50*, 891–905. [CrossRef]
15. Brown, B.D.; Card, H.C. Stochastic Neural Computation II: Soft Competitive Learning. *IEEE Trans. Comput.* **2002**, *50*, 906–920. [CrossRef]
16. Ardakani, A.; Leduc-Primeau, F.; Onizawa, N.; Hanyu, T.; Gross, W.J. VLSI Implementation of Deep Neural Network Using Integral Stochastic Computing. *IEEE Trans. Very Large Scale Integr. (VLSI) Syst.* **2017**, *25*, 2688–2699. [CrossRef]
17. Liu, S.; Jiang, H.; Liu, L.; Han, J. Gradient Descent Using Stochastic Circuits for Efficient Training of Learning Machines. *IEEE Trans. Comput.-Aided Des. Integr. Circuits Syst.* **2018**, *37*, 2530–2541. [CrossRef]
18. Liu, Y.; Liu, L.; Lombardi, F.; Han, J. An Energy-Efficient and Noise-Tolerant Recurrent Neural Network Using Stochastic Computing. *IEEE Trans. Very Large Scale Integr. (VLSI) Syst.* **2019**, *27*, 2213–2221. [CrossRef]
19. Liu, Y.; Liu, S.; Wang, Y.; Lombardi, F.; Han, J. A Survey of Stochastic Computing Neural Networks for Machine Learning Applications. *IEEE Trans. Neural Netw. Learn. Syst.* **2020**, *32*, 2809–2824. [CrossRef]
20. Temenos, N.; Sotiriadis, P.P. Nonscaling Adders and Subtracters for Stochastic Computing Using Markov Chains. *IEEE Trans. Very Large Scale Integr. Syst. (VLSI)* **2021**, *29*, 1612–1623. [CrossRef]
21. Li, P.; Lilja, D.J. Using Stochastic Computing to Implement Digital Image Processing Algorithms. In Proceedings of the IEEE 29th International Conference on Computer Design (ICCD), Amherst, MA, USA, 9–12 October 2011.
22. Li, P.; Lilja, D.J.; Qian, W.; Bazargan, K.; Riedel, M.D. Computation on Stochastic Bit Streams Digital Image Processing Case Studies. *IEEE Trans. Very Large Scale Integr. Syst. (VLSI)* **2014**, *2*, 449–462. [CrossRef]
23. Alaghi, A.; Li, C.; Hayes, J.P. Stochastic circuits for real-time image-processing applications. In Proceedings of the IEEE 50th ACM/EDAC/IEEE Design Automation Conference (DAC), Austin, TX, USA, 29 May–7 June 2013.
24. Temenos, N.; Sotiriadis, P.P. Stochastic Computing Max & Min Architectures Using Markov Chains: Design, Analysis, and Implementation. *IEEE Trans. Very Large Scale Integr. Syst. (VLSI)* **2021**, *29*, 1813–1823.
25. Alawad, M.; Lin, M. Fir filter based on stochastic computing with reconfigurable digital fabric. In Proceedings of the IEEE 23rd Annual International Symposium on Field-Programmable Custom Computing Machines, Vancouver, BC, Canada, 2–6 May 2015.
26. Vlachos, A.; Temenos, N.; Sotiriadis, P.P. Exploring the Effectiveness of Sigma-Delta Modulators in Stochastic Computing-Based FIR Filtering. In Proceedings of the IEEE 10th International Conference on Modern Circuits and Systems Technologies (MOCAST), Thessaloniki, Greece, 5–7 July 2021.

27. Saraf, N.; Bazargan, K.; Lilja, D.J.; Riedel, M.D. IIR filters using stochastic arithmetic. In Proceedings of the IEEE Design, Automation & Test in Europe Conference & Exhibition (DATE), Dresden, Germany, 24–28 March 2014.
28. Ahmed, K.J.; Yuan, B.; Lee, M.J. High-Accuracy Stochastic Computing-Based FIR Filter Design. In Proceedings of the IEEE International Conference on Acoustics, Speech and Signal Processing (ICASSP), Calgary, AB, Canada, 15–20 April 2018.
29. Ichihara, H.; Sugino, T.; Ishii, S.; Iwagaki, T.; Inoue, T. Compact and Accurate Digital Filters Based on Stochastic Computing. *IEEE Trans. Emerg. Top. Comput.* **2019**, *7*, 31–43. [CrossRef]
30. Liu, Y.; Parhi, K.K. Architectures for Recursive Digital Filters Using Stochastic Computing. *IEEE Trans. Signal Process.* **2016**, *64*, 3705–3718. [CrossRef]
31. Tehrani, S.S.; Gross, W.J.; Mannor, S. Stochastic decoding of LDPC codes. *IEEE Commun. Lett.* **2006**, *10*, 716–718. [CrossRef]
32. Basetas, C.; Temenos, N.; Sotiriadis, P.P. Comparison of Recently Developed Single-Bit All-Digital Frequency Synthesizers in Terms of Hardware Complexity and Performance. In Proceedings of the IEEE International Symposium on Circuits and Systems (ISCAS), Florence, Italy, 27–30 May 2018.
33. Basetas, C.; Temenos, N.; Sotiriadis, P.P. An Efficient Hardware Architecture for the Implementation of Multi-Step Look-Ahead Sigma-Delta Modulators. In Proceedings of the IEEE International Symposium on Circuits and Systems (ISCAS), Florence, Italy, 27–30 May 2018.
34. Schreier, R.; Temes, G.C. *Understanding Delta-Sigma Data Converters*; John Wiley & Sons, Inc.: Hoboken, NJ, USA, 2005.
35. Liu, S.; Han, J. Toward Energy-Efficient Stochastic Circuits Using Parallel Sobol Sequences. *IEEE Trans. Very Large Scale Integr. (VLSI) Syst.* **2018**, *26*, 1326–1339. [CrossRef]
36. Qian, W.; Li, X.; Riedel, M.D.; Bazargan, K.; Lilja, D. An Architecture for Fault-Tolerant Computation with Stochastic Logic. *IEEE Trans. Comput.* **2011**, *60*, 93–105. [CrossRef]
37. Camps, O.; Stavrinides, S.; Picos, R. Stochastic Computing Implementation of Chaotic Systems. *Mathematics* **2021**, *9*, 375. [CrossRef]
38. Lee, V.T.; Alaghi, A.; Ceze, L. Correlation manipulating circuits for stochastic computing. In Proceedings of the IEEE Design, Automation & Test in Europe Conference & Exhibition (DATE), Dresden, Germany, 19–23 March 2018.
39. Alaghi, A.; Hayes, J.P. Exploiting correlation in stochastic circuit design. In Proceedings of the IEEE 31st International Conference on Computer Design (ICCD), Asheville, NC, USA, 6–9 October 2013.
40. Liu, Y.; Parhi, M.; Riedel, M.D.; Parhi, K.K. Synthesis of Correlated Bit Streams for Stochastic Computing. In Proceedings of the IEEE 50th Asilomar Conference on Signals, Systems and Computers, Pacific Grove, CA, USA, 6–9 November 2016.
41. Chang, Y.; Parhi, K.K. Architectures for digital filters using stochastic computing. In Proceedings of the IEEE International Conference on Acoustics, Speech and Signal Processing, Vancouver, BC, Canada, 26–31 May 2013.
42. Meyer-Baese, U. *Digital Signal Processing with Field Programmable Gate Arrays*, 4th ed.; Springer: Berlin,Heidelberg, Germany, 2014.
43. Ren, A.; Li, Z.; Ding, C.; Qiu, Q.; Wang, Y.; Li, K.; Qian, X.; Yuan, B. SC-DCNN: Highly-Scalable Deep Convolutional Neural Network using Stochastic Computing. In Proceedings of the ACM 22nd International Conference on Architectural Support for Programming Languages and Operating Systems(ASPLOS), Xi'an, China, 8–12 April 2017.
44. Ichihara, H.; Ishii, S.; Sunamori, D.; Iwagaki, T.; Inoue, T. Compact and Accurate Stochastic Circuits with Shared Random Number Sources. In Proceedings of the IEEE 32nd International Conference on Computer Design (ICCD), Seoul, Korea, 19–22 October 2014.
45. Stine, J.E.; Castellanos, I.; Wood, M.; Henson, J.; Love, F.; Davis, W.R.; Franzon, P.D.; Bucher, M.; Basavarajaiah, S.; Oh, J.; et al. FreePDK: An Open-Source Variation-Aware Design Kit. In Proceedings of the IEEE International Conference on Microelectronic Systems Education, San Diego, CA, USA, 3–4 June 2007.

MDPI

Article

A Simplified Tantalum Oxide Memristor Model, Parameters Estimation and Application in Memory Crossbars [†]

Valeri Mladenov * and **Stoyan Kirilov**

Department of Fundamentals of Electrical Engineering, Technical University of Sofia, 1000 Sofia, Bulgaria; s_kirilov@tu-sofia.bg
* Correspondence: valerim@tu-sofia.bg; Tel.: +359-965-2131
† This paper is an extended version of the paper published in 10th International Conference on Modern Circuit and System Technologies on Electronics and Communications (MOCAST 2021) IEEE Proceeding, Thessaloniki, Greece, 5–7 July 2021.

Abstract: In this paper, an improved and simplified modification of a tantalum oxide memristor model is presented. The proposed model is applied and analyzed in hybrid and passive memory crossbars in LTSPICE environment and is based on the standard Ta_2O_5 memristor model proposed by Hewlett–Packard. The discussed modified model has several main enhancements—inclusion of a simplified window function, improvement of its effectiveness by the use of a simple expression for the i–v relationship, and replacement of the classical Heaviside step function with a differentiable and flat step-like function. The optimal values of coefficients of the tantalum oxide memristor model are derived by comparison of experimental current–voltage relationships and by using a procedure for parameter estimation. A simplified LTSPICE library model, correspondent to the analyzed tantalum oxide memristor, is created in accordance with the considered mathematical model. The improved and altered Ta_2O_5 memristor model is tested and simulated in hybrid and passive memory crossbars for a state near to a hard-switching operation. After a comparison of several of the best existing memristor models, the main pros of the proposed memristor model are highlighted—its improved implementation, better operating rate, and good switching properties.

Keywords: memristor model; tantalum oxide; memory crossbar; nonlinear dopant drift; window function; LTSPICE library memristor model

Citation: Mladenov, V.; Kirilov, S. A Simplified Tantalum Oxide Memristor Model, Parameters Estimation and Application in Memory Crossbars. *Technologies* **2022**, *10*, 6. https://doi.org/10.3390/technologies10010006

Academic Editors: Spiros Nikolaidis and Rodrigo Picos

Received: 26 November 2021
Accepted: 5 January 2022
Published: 10 January 2022

Publisher's Note: MDPI stays neutral with regard to jurisdictional claims in published maps and institutional affiliations.

1. Introduction

The memristor elements, based mainly on amorphous transition metal oxides, as TiO_2, HfO_2, Ta_2O_5, ZnO and other materials, have many potential applications, as in nonvolatile memory devices, artificial neural networks, and logical and reconfigurable electronic circuits [1–4]. The memristor was predicted in 1971 by Chua as the fourth fundamental, passive, and nonlinear one-port electronic element [5]. It has a memory effect and can retain its state after switching off the electric sources [5,6]. The first physical prototype of a memristor, based on TiO_2, was proposed by Hewlett–Packard (HP) Research Team, supervised by Stanley Williams [6]. In the scientific literature, information for generated polymeric, ferroelectric, and other types of memristors has been published [7–10]. Along with the oxide materials with resistance switching abilities already analyzed, the amorphous Ta_2O_5, doped by oxygen vacancies, has excellent switching properties, a sound dependability, great operating speed, small energy consumption, comparatively extensive memorizing time, and a good compatibility with the commonly used Complementary Metal Oxide Semiconductor (CMOS)-integrated circuits technology [11–14]. Due to this reason and to the increased interest of engineers and scientists to tantalum oxide memristors in the recent years, these elements and their promising applications are a central object of analysis in the paper. The tantalum oxide-based memory elements involve a high-conducting region and a channel, based on non-stochiometric Ta$_2$O$_5$ amorphous material [15–18]. The memristance (abbreviated from memory resistance) and the correspondent memristor status might be changed by applying outer voltage or current pulses [19–21].

For an accurate description of the tantalum oxide memristors' behavior in electronic digital and analog circuit and devices, a precise and simplified corresponding model is needed. Many attempts for adapting to the widely spread TiO$_2$ memristor models for approximate representation of the considered Ta$_2$O$_5$ memristors are available in the scientific literature [11]. Conversely, the structure and principal of operation of TiO$_2$ and Ta$_2$O$_5$ memristor elements are distinct [6,11]. This is the major cause for the generation of various special tantalum oxide memristor models [11,13,14]. The classical memristor model [11] suggested by Hewlett–Packard Research Labs has a high accuracy and sound switching representation. It applies the standard Heaviside step expression in the differential state equation and a non-differentiable modulus expression in the respective *i–v* relationship. Regrettably, these relationships are not flat and differentiable [11,13], and when this mathematical model is used for Simulation Program with Integrated Circuit Emphasis (SPICE) memristor model generation, many problems with the convergence exist [22,23]. This is a disadvantage of the traditional Ta$_2$O$_5$ memristor model. An improvement of the standard model is presented in [13]. The modified tantalum oxide memristor model [13] is related to a good precision and an enhanced behavior in SPICE environment. It uses continuous and differentiable relations in describing equations as an alternative of the applied non-smooth and non-differentiable expressions in [11]. While the enhanced tantalum oxide-based memristor model [13] is appropriate for SPICE incorporation, it is a computationally complex and time utilizing one, owing to the applied modulus-like complex expression. It demands many elementary calculations, corresponding to the high number of exponents in the state equation [13,14]. No window has been included in this model, and sometimes, the state variable *x* might go outside the interval between 0 and 1 when the memristor operates in a hard-switching mode. This is one more disadvantage of the discussed memristor models. A distinct memristor model, presented in [16], is appropriate for Ta$_2$O$_5$ memristor representation. It has good accurateness and correctly represents the switching processes in the memristor element. Disappointingly, it is founded on the classical non-differentiable step function, which is associated with convergence issues [20]. The memristor model represented in [15] is relatively simple and accurate, with a high operating rate.

The motivation for the present paper is the partial absence of simple and accurate Ta$_2$O$_5$ memristor models. The purpose of this research is to propose and consider an accurate, adjustable, fast operating, and simple model [12] for Ta$_2$O$_5$ memristors appropriate for generation of corresponding Linear Technology SPICE (LTSPICE) [23] library model. LTSPICE is a simple, user-friendly, and free software for analysis and design of electronic circuits and devices by preliminary computer simulations [23]. This software is offered by Analog Devices Corporation and could be freely downloaded and installed, using the next link: https://www.analog.com/en/design-center/design-tools-and-calculators/ltspice-simulator.html (last accessed on 5 November 2021). LTSPICE software is preferrable for electronic device simulations by many design engineers and scientists. For adjustment of the offered memristor model, experimental current-voltage characteristics of tantalum oxide memristors, results derived from the use of several of the best existing models [11,13,14], and different methods for optimal parameters' estimation [22,24] are applied. The parameters' estimation procedure is realized in the MATLAB environment [22]. The least mean square error (MSE) between the experimental and simulated voltage–current relationships is used as an optimization criterion [22,24,25]. A corresponding LTSPICE library memristor model is successfully created. It is included in a unified and open LTSPICE memristor library, freely available for download and use at: https://github.com/mladenovvaleri/Advanced-Memristor-Modeling-in-LTSpise (last accessed on 9 November 2021) [26]. The proposed memristor model is analyzed and tested in passive and hybrid memristor memory crossbars for reading, writing, and erasing processes [27–30]. The derived current–voltage and state-flux relationships confirm the proper operation of the considered tantalum oxide-based memristor model for soft-switching and hard-switching modes.

The rest of this paper is organized as follows. A description of the basic tantalum oxide memristor models is shown in the next section. The adjustment of the offered memristor model using experimental *i–v* characteristics, a procedure for extraction of the optimal model's parameters, and a procedure for coefficients' assessment in MATLAB—Simulink environment are discussed in Section 3. The related LTSPICE memristor model and its analysis are described in Section 4. The operation of the commented LTSPICE memristor library model in passive and hybrid memristor crossbars is shown in Section 5. The results are discussed in Section 6. The conclusion is given in Section 7.

2. The Basic Existing Ta_2O_5 Memristor Models and the Proposed Modification

For better understanding and completeness of the models' description, the basic existing standard and modified Ta_2O_5 memristor models are briefly discussed in this section. The tantalum oxide memristor element has two terminals—top electrode (TE), also known as anode, and bottom electrode (BE), namely the cathode [11,18]. These electrodes are made of platinum or tantalum-layered materials [11]. Several modified physical realizations of tantalum oxide-based memristors with layered structure are available [19]. This two-terminal passive electronic element has a cross-section with a square shape. The injection of oxygen vacancies is conducted by electroforming-like process at comparatively low-level voltage pulses [11]. In the Ta_2O_5 memristor nanostructure, several parallel-oriented conducting channels are available [18,19]. The peripheral region is based on pure and stochiometric Ta_2O_5 amorphous material [11]. The central conducting channel is founded on a consistent solution of atoms in a crystalline a material. A transitional region, made of a non-stochiometric Ta_2O_5 and partially doped with O_2 vacancies, is created among the central and the peripheral channels of the tantalum oxide memristor element [11,19]. The memristor status could be changed by the use of outer voltage pulses. According to several physical factors, as the effective ionic dopant mobility, the memristor length and the resistance in the operating state, the memristor could function at different frequencies and amplitudes of the applied signals, representing the change of the memristance and the correspondent state variable. The following tantalum oxide memristor models are able to properly represent the memristor behavior for low, middle and higher frequencies for hard- and soft-switching modes. The memristor state variable x [11,13] is stated as a ratio of the areas of the intersections of the low-resistance channel, denoted by a_1, and the area of the whole nanostructure, expressed by a_2 [11]:

$$x = \frac{a_1}{a_2} \qquad (1)$$

According to physical considerations $(a_1 \leq a_2)$, the memristor state variable x is limited in the range $(0, 1)$. The described physical limitation of the memory element's state variable for the memristor models could be mathematically realized, using an appropriate window function [7,8,28]. If a voltage signal with comparatively low level and high frequency is applied to the memristor, then its state variable x does not reach its limiting values. This operation is also known as a soft-switching mode. In such a case, the correspondent current–voltage relationship is a multi–valued pinched hysteresis loop, and the respective state-flux relation is a single-valued curve. For higher-level and lower-frequency signals, the state variable x could reach its boundary values—zero and unity. In this case, the memristor operates in the so-called hard-switching mode. This functioning is related to a rectifying effect [7,8,31]. The terminal-state problems are related to the impossibility of the state variable x to be changed when it reaches its limiting values. Some window functions, as these proposed by Joglekar and Williams [6,31], are related to terminal state problems, while others as Biolek's [7] are able to correctly resolve these problems. In the present case, terminal state issues are not established, owing to the use of a modified Biolek-based window function [12,26].

2.1. The Standard Model of Ta$_2$O$_5$ Memristor, Proposed by HP

The highest value of the *memductance* (abbreviated from *memristor conductance*) of the Ta$_2$O$_5$ memristor is denoted by G_{max} [11,13]. The conductivity of the doped layer of the memristor could be represented by the Frenkel–Poole equation [11,13]. According to the traditional HP model of tantalum oxide memristors [11], the state-dependent current–voltage relationship could be described as:

$$i(v,x) = v \cdot G_{eq} = v \cdot \left[x \cdot G_{\max} + a \cdot (1-x) \cdot \exp\left(b \cdot \sqrt{|v|} \right) \right] \tag{2}$$

where i is the memristor current, v is the applied voltage across the memristor, G_{eq} is the equivalent memductance, and a and b are tuning parameters. The state differential equation of the standard tantalum oxide memristor has the following expression [11]:

$$\begin{aligned}
\frac{dx}{dt} = {}& A \cdot \sinh\left(\frac{v}{\sigma_{OFF}} \right) \cdot \exp\left(\frac{1}{\beta\, v\, i+1} - \frac{x_{OFF}^2}{x^2} \right) \cdot stp(-v) + \\
& + B \cdot \sinh\left(\frac{v}{\sigma_{ON}} \right) \cdot \exp\left(\frac{v\, i}{\sigma_P} - \frac{x^2}{x_{ON}^2} \right) \cdot stp(v)
\end{aligned} \tag{3}$$

where A, B, σ_{OFF}, σ_{ON}, σ_P, x_{ON}, β and x_{OFF} are coefficients for adjustment of the model [11,13,31]. The memristor model is based on a nonlinear dopant drift representation [11]. The applied in the model Heaviside step function $stp(\cdot)$ is [7,8,25]:

$$\left| \begin{aligned}
stp(v) &= 1,\ v \in [0,\ +\infty) \\
stp(v) &= 0,\ v \in (-\infty,\ 0)
\end{aligned} \right. \tag{4}$$

The classical HP tantalum oxide memristor model [11] is completely characterized by (1) and (2). This model holds good accuracy, related to an RMS error of about 2.86% [11,13]. It might correctly describe the performance of memristors, based on amorphous tantalum oxide in electronic schemes and devices [11,13]. The main drawbacks of the model, related to SPICE realization, are the application of an interrupted and non-differentiable step function and a non-smooth and non-differentiable modulus function [11,13]. The analysis of the memristor model is realized by computer simulations in MATLAB environment [22]. The simulations are prepared on a desktop computer system with Intel *i5*, 2.4 GHz 4-core microprocessor, Windows 10 system, and 8 GB Random Access Memory (RAM) [12]. The time for simulation of this memristor model is t_1 = 49.5 ms.

2.2. Enhancements of the Classical HP Memristor Model

The major improvement included in [13], which enhances the SPICE realization of the memristor model, is the replacement of the classical modulus expression and the standard step function by differentiable and flat replacements [25]. A simple step-like differentiable relation, applied in [13,14], is:

$$stpp(v) = [\exp(k\, v) + 1]^{-1} \tag{5}$$

where k is a tuning coefficient [13,25]. It determines the steepness of the step-like function $stpp(.)$ in the region of switching [25]. For SPICE realization, it usually has a value between -50 and -1000 [13,14]. A differentiable and flat analog of the classical modulus function, used in (2) is [13,14]:

$$f_m(v) = v \cdot \left[\frac{1}{\exp(-\rho\, v) + 1} - \frac{1}{\exp(\rho\, v) + 1} \right] \tag{6}$$

where ρ is a parameter for fitting the modulus function [13]. Usually, for SPICE realization, its value is between 100 and 1000 [13,14]. The main improvement in this memristor model is the realized prevention of convergence difficulties in SPICE environment [13]. This model is with good accuracy, and the related error is about 2.88%. A disadvantage of the described memristor model is its comparatively high computational complicacy. The necessary time for simulation of the considered tantalum oxide memristor model is $t_2 = 16.8$ ms.

2.3. The Proposed and Considered Improved Tantalum Oxide Memristor Model

The proposed modified memristor model [12] contains several main replacements. First, the classical step expression $stp(v)$ in (3) is substituted by a smooth and differentiable analogue $s(v)$ [12,25]:

$$s(v) = \frac{1}{2}\left(1 + \frac{v}{\sqrt{v^2 + m}}\right) \tag{7}$$

where m is a coefficient with a typical value between 0.01 and 0.0001 [12,25]. This coefficient determines the steepness of the step-like function in the region of switching [25]. The function expressed by (7) is an alternative of (5). An advantage of such a step-like continuous function is the partial avoidance of convergence problems in SPICE environment. Conversely, the correspondent region of switching is not as sharp and rigorously defined, as in the case of the classical Heaviside function [25]. This could lead to decreasing the accuracy of the respective memristor model, especially if it operates at low-voltage signals and soft-switching mode. A compromise between the model's accuracy and the prevention of convergence issues must be introduced, and the coefficient m could be used as a parameter for adjustment of the memristor model. Due to the use of simple mathematical expressions and avoiding the exponential function applied in (5), the memristor model based on (7) has a slightly higher operating speed [12]. The operating rate is related to the number of the elementary mathematical operations in the memristor model, when it is functioning in a software environment, as MATLAB [22] or SPICE [23]. Second, the fragment $a \cdot \exp\left(b \cdot \sqrt{|v|}\right)$ in (2) is approximated with a low-order polynomial:

$$F(v) = a \cdot \exp\left(\sqrt{|v|} \cdot b\right) \approx h_1\, v^4 + h_2\, v^2 + h_3 \tag{8}$$

which holds almost equal values in the interval $(-1\text{ V}, 0.5\text{ V})$. The RMS error among the original term and its approximation in (8) is about 2.8% [12]. The coefficients h_1, h_2 and h_3 in the right-hand position of (8) are parameters for adjustment of the polynomial [12,22]. The applied approximation in (8) ensures lower complexity of the considered tantalum oxide memristor model, compared to the model described in [11]. Third, a modified and comparatively simple window representation $f_{Bmod}(x,i)$, founded on both the classical window expression proposed by Biolek [7] and the described step-like function $s(i)$, is used in the considered memristor model [12]:

$$f_{Bmod}(x,\, i) = 1 - [s(-i) - x]^2 \tag{9}$$

The applied window function is able not only to restrict the state variable x in the interval $(0, 1)$ but also to correctly represent the boundary effects and to solve the terminal state problems [7,12]. The terminal state problems are related to several memristor models, as Strukov–Williams, Joglekar and others [8]. Sometimes, when the state variable reaches the physical limits of zero and unity, it cannot be changed, although the applied voltage and the correspondent flux linkage are with sufficient values and polarities. Other models, such as Biolek and Boundary Condition Memristor (BCM) models, resolve successfully this issue [8]. The term $s(-i)$ is a smooth and differentiable step-like function, used for prevention of convergence issues in SPICE environment [13,14]. The memristor state differential Equation (3) [12] of the described model might be expressed in the next form:

$$\frac{dx}{dt} = [M(v)E(i,x,v)s(-v) + C(v)J(i,v)s(v)]f_{Bmod}(\,i,x) \tag{10}$$

where the terms $M(v)$, $E(x,i,v)$, $C(v)$, $J(x,i,v)$, are expressed according to (3) as follows:

$$M(v) = A \cdot \sinh\left(\frac{v}{\sigma_{OFF}}\right) \tag{11}$$

Here, A and σ_{OFF} are fitting parameters [11,13]. Unfortunately, the steepness of this term as a function of v is high in some voltage ranges and cannot be correctly approximated, unlike the term $C(v)$. The next term in (10) $E(x,i,v)$ is expressed by (12):

$$E(x,i,v) = \exp\left(\frac{1}{\beta\, v\, i + 1} - \frac{x_{OFF}^2}{x^2}\right) \tag{12}$$

where β and x_{OFF} are parameters for adjustment of the memristor model [13]. The term (12) is also not appropriate for approximation and simplification [13]. The term $C(v)$ is successfully approximated by a low-order polynomial [12]:

$$C(v) = B \cdot \sinh\left(\frac{v}{\sigma_{ON}}\right) \approx k_1\, v^3 + k_2\, v \tag{13}$$

where B, σ_{ON}, k_1, k_2 are fitting coefficients [12]. The coefficients k_1 and k_2 are derived using the least squares approximation method, realized in MATLAB environment [22]. The RMS error between the original and the approximated values of $C(v)$ in the interval $(-1.45$ V, 1.45 V$)$ is about 2.4% [12]. The next term $J(x,i,v)$ is [11–13]:

$$J(x,i,v) = \exp\left(\frac{v\, i}{\sigma_P} - \frac{x^2}{x_{ON}^2}\right) \tag{14}$$

Here, σ_P, x_{ON} are fitting parameters [11,13]. The final term in the state equation, presented by (14) could not be appropriately simplified. The original and the approximated expressions of the memristor current i as a function of the applied voltage v and the equivalent memductance of the tantalum oxide memristor $G_{eq}(x,v)$ are [12]:

$$\begin{aligned}
i(x,v) = v \cdot G_{eq} &= v \cdot \left[x \cdot G_{max} + a \cdot (1 - x) \cdot \exp\left(b \cdot \sqrt{|v|}\right) \right] = \\
&= v \cdot \left[x \cdot G_{max} + \left(h_1\, v^4 + h_2\, v^2 + h_3\right) \cdot (1 - x)\right]
\end{aligned} \tag{15}$$

The proposed simplified and improved tantalum oxide memristor model is completely described by (10) and (15) [12]. The model has good precision, and the obtained error is about 2.93%. It is able to properly operate in high-frequency mode, representing the respective alteration of the state variable x for both soft-switching and hard-switching operation [12].

3. The Fine Tuning and Parameters Estimation of the Suggested Memristor Model

The enhanced tantalum oxide-based memristor model [12], considered in this paper and fully described by (10) and (15), includes several parameters for fine adjustment. It is tuned in accordance with voltage–current characteristics, obtained by experimental data [11,19], the original HP model [11] and several of the best and accurate models [11,13,14,19], and it employs a procedure for changing the model's factors until reaching the total minimum of the RMS error [12,22]. Several researchers [31,32] have applied simulation annealing and gradient descent techniques for deriving the best parameters of the optimized memristor models. Additional comparison to several of the best tantalum oxide memristor models [11,13,14] is conducted, applying current–voltage and the correspondent state-flux relationships. Considering the terms (8), (13), and (10), (15), it could be easily concluded that the evolution of the memristor state variable x depends mainly on its initial value x_0 and the history of the applied voltage signal. The retention of the memristor state and the related information stored in the element are dependent on the state variable and the corresponding memductance. The processes of internal diffusion of dopant ions could affect the memristor state, but over a long interval, practically about ten years [11,19]. Owing to this, additional terms for considering the reliability and retention [33] related to the tantalum oxide memristor are not included; thus, the model must be simple and applicable for analysis of memristor-based schemes and devices. A technique for extraction of the memristor model's parameters in the MATLAB environment [22] is also applied. The considered method for optimization of the tantalum oxide memristor model's tuning is based on altering the coefficients and searching for the global minimum of the RMS error among the experimental and the obtained-by-simulations $i–v$ relationships. At each iteration, one of the parameters of the memristor model is changing by a little constant increment [12]. The corresponding RMS error between the experimental and the simulated current–voltage characteristics is calculated. The other coefficients for tuning the memristor model are also altering. After finishing the course of this tuning process, a graphical observation of the simulated $i–v$ relation and its closeness to the experimental current–voltage characteristic is also established [12], paying attention to the shape of the obtained pinch hysteresis loop and especially its regions of switching the memristance [12]. The corresponding time diagrams of the experimental and the simulated memristor currents are compared as well, applying the squared differences between them in the sampling points.

The decisive factor for finishing the tuning procedure is the minimization of the RMS error [12,22,24]. Supplementary tests and simulations are made in the proximity of the derived optimal levels of the tantalum oxide memristor model's coefficients, using decreased increments for their changing [12]. The precise and optimal levels of the memristor element model's coefficients are also established, applying the least squares method in Simulink environment, using the Optimization Toolbox [22]. The technique for memristor model's parameters estimation could be approximately summarized in several basic steps:

1. Initialization of the values of the model's parameters, using the coefficients in the original sources [11,13,19] and the coefficients in the polynomials (8), (13) after their approximation in MATLAB environment [22];
2. Defining the satisfactory root mean square error for stopping the procedure, the respective tolerances and the maximal iteration steps;
3. Starting simulation and calculating the values of the memristor current, according to the applied model;
4. Estimation of the root mean square error (the cost function) between the experimental and the calculated memristor currents;
5. Changing the model's parameters, according to the gradient descent of the cost function, then proceed to step 4;
6. If the satisfactory root mean square error is reached, or the maximal iteration steps are finished, stopping the simulation and estimation of the obtained model's parameters; if not, proceed to step 5.

The applied voltage signals in the Simulink model of the memristor are hitherto sampled, and the time step is 10 µs. The voltage drop across the element v_1 is applied as an input signal, applied to the Simulink model of the considered memristor [12,22]. The empirically recorded memristor's current is denoted by i_{mes}. The simulated output of the Simulink memristor model is equivalent to the simulated current of the tantalum oxide memory element i_{calc}. The cost function S_{cost} is represented by a sum of the squares of the differences of the calculated and experimental memristor current's values [12,22]:

$$S_{\cos t} = \sum_{k=1}^{N} [i_{calc}(k) - i_{mes}(k)]^2 \qquad (16)$$

where N = 200,000 is the total amount of samples of the considered signals, and k is the actual sample. The benchmark for finishing the parameters' estimation procedure minimizes the cost function S_{cost} [22,32]. The best possible values of the extracted memristor model's coefficients are presented in Table 1 for the generation of the LTSPICE tantalum oxide memristor library model, as discussed in the next section. The derived optimal values of the tantalum oxide memristor model's coefficients are similar to those obtained in other scientific works on tantalum oxide memristors [11,13,14]. The obtained values of the coefficients k_1, k_2, h_1, h_2 and h_3 after approximations in MATLAB environment [22] ensure that the RMS error between the respective terms is lower than 0.8%.

Table 1. Memristor model parameters derived after the optimization procedures.

Memristor Parameters	A	σ_{OFF}	β	x_{OFF}	k_1	k_2	σ_P
Units	s^{-1}	V	W^{-1}	-	$s^{-1}V^{-3}$	$s^{-1}V^{-1}$	W
Optimal values	1.47×10^{-7}	0.042	476.5	0.27	-1.3×10^{-7}	0.0001	7.05×10^{-5}
Memristor parameters	x_{ON}	m	x_0	G_{max}	h_1	h_2	h_3
Units	-	V^2	-	S	SV^{-4}	SV^{-2}	S
Optimal values	0.04	6.2×10^{-8}	0.1	0.0227	-1.98×10^{-6}	1.35×10^{-4}	3.31×10^{-4}

The scaled values of the trajectories of the model's parameters during the optimization are shown in Figure 1a to present their change with time. The initial values of the model's parameters are chosen to be close to those obtained in [11,13], and the memristor state variable x is in the space (0, 1).

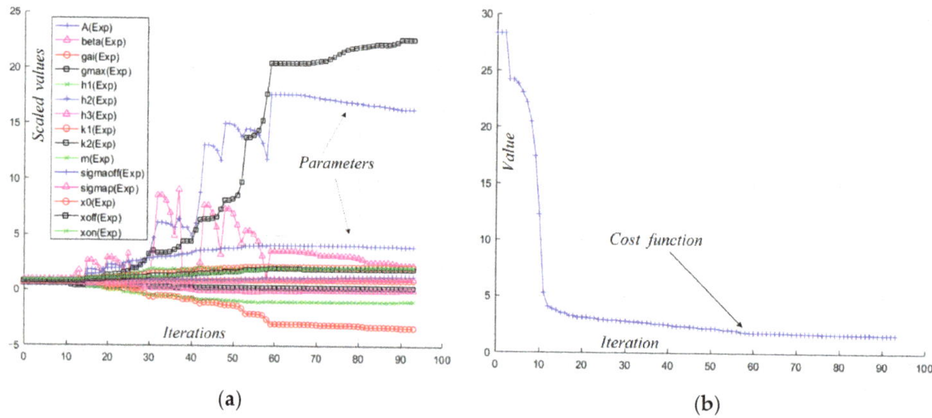

(a)   **(b)**

Figure 1. (**a**) Trajectories of the memristor model parameters during the optimization procedure, expressed in scaled values. (**b**) Minimization of the cost function.

For a brief comparison of the presented model to some of the existing and frequently used tantalum oxide memristor models [11,13–15], several basic criteria, as simulation time, operation speed, accuracy and convergence are presented in Table 2.

Table 2. A brief comparison of the presented model to several of the best Ta_2O_5 memristor models.

Models	Accuracy	Simulation Time, ms	Operating Speed	Convergence Issues
[11]	very high	49.52	average	rarely observed
[13]	high	16.84	high	not observed
[14]	high	16.42	high	not observed
[12]	high	16.31	high	not observed

The minimization of the cost function S_{cost} is given in Figure 1b to show the correctness of the optimization process [12]. The time diagrams of the memristor current (experimental and simulated) and voltage after the optimization are presented in Figure 2a for confirmation of the sufficient closeness between them. The experimental current–voltage characteristic [11] and the simulated i–v relationship obtained in the MATLAB environment [22] are presented in Figure 2b to present the proximity between the characteristics.

The obtained RMS error is about 3.24%. The time for simulation of the memristor model t_3 is about 16.3 ms [12].

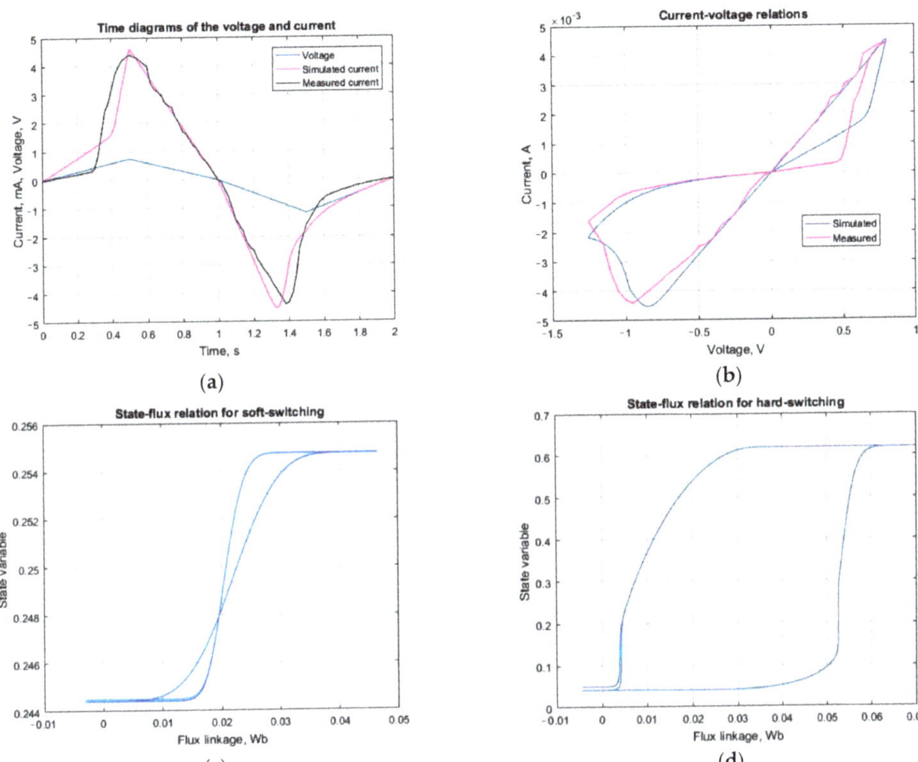

Figure 2. (**a**) Time charts of the memristor voltage and current (experimental and simulated) after the optimization process. (**b**) Current–voltage relationships obtained after the parameters estimation procedure. (**c**) State-flux relation for soft-switching; (**d**) State-flux relation for hard-switching.

Owing to the modified and simple equations, the suggested model of Ta_2O_5 memristor operates more rapidly than the formerly expressed memristor models [11,13,14]. The derived state-flux relationships for soft-switching and hard-switching operations are presented in Figure 2c,d, respectively, for confirmation of the proper functioning of the suggested modified memristor model [12]. The state-flux characteristic of the tantalum oxide memristor for soft-switching mode is a pinched hysteresis loop, while for the hard-switching operation, it is a hysteresis curve, correspondent to the memristor boundary effects [31].

4. The Corresponding LTSPICE Memristor Library Model

Based on the proposed improved mathematical memristor model, described by (10) and (15), an LTSPICE [23] library model of the described tantalum oxide-based memristor element was created [12,26]. The basic functional units in the LTSPICE environment are employed for realization of the relevant math calculus, in accordance with the presented memristor model. The substituting schematic of the generated LTSPICE model is shown in Figure 3a for further discussion. The memristor variable x is realized as the voltage $V(Y)$ of the capacitive element C_1 [12,26]. Its current corresponds to the time derivative of x. The two-port voltage controlled current source G_1 represents the memductance G of the tantalum oxide memory element. The internal own resistance of the applied voltage source V_1 is represented by the resistance R_2. The resistor R_1 prevents the occurrence of convergence issues [12,13]. The generated LTSPICE memristor library model in a simple electric circuit is presented in Figure 3b. The main terminals are the top electrode (TE, anode) and the bottom electrode (BE, cathode). The additional electrode Y is applied for measuring the memristor state variable x [12,13].

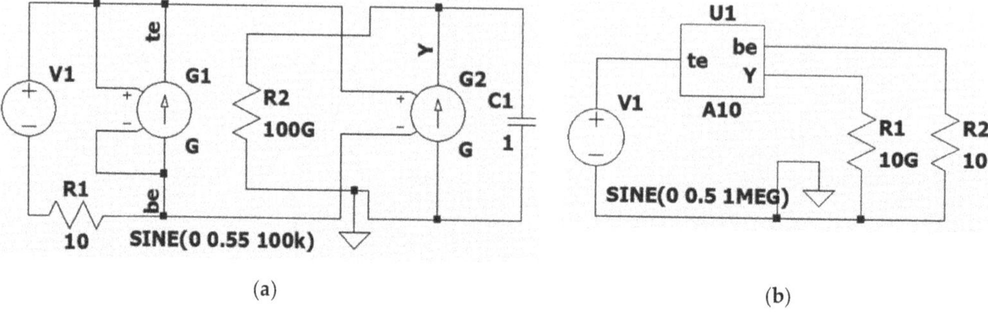

(a)   (b)

Figure 3. (a) An equivalent LTSPICE schematic of the considered tantalum oxide-based memristor model. (b) The generated LTSPICE memristor model, included in a simple test electric circuit.

The correspondent LTSPICE code of the memristor model is [12,26]:
.subckt *A10 a c Y*
*terminals—top electrode (a), bottom electrode (c) and additional
*electrode Y for measuring the memristor state variable
.params y_{on} = 0.04 A = 1.37 × 10^{-7} *sigmap* = 7.05 × 10^{-5} *sigmaoff* = 0.042
.params G_m = 0.027 y_{off} = 0.27 m = 6 × 10^{-10} *beta* = 822.6 k_1 = 0.0062
.params k_2 = 0.0001 h_1 = 1.98 × 10^{-4} h_2 = 0.000135 h_3 = 3.31 × 10^{-4}
*memristor model parameters for tuning
G_1 *a c* value = {($V(Y)$ × G_m +(1 − $V(Y)$) × (h_1 × (pow($V(a,c)$,4)) + h_2 × (pow($V(a,c)$,2)) + h_3)) × $V(a,c)$}
*voltage-controlled current source G_1 for deriving the memristor current
G_2 *Y* 0 value = {(A × sinh($V(a,c)$/*sigmaoff*) × exp(1/(1 + *beta* × $I(G_1)$ × exp(−pow(y_{off}/$V(Y)$,2)) × $V(a,c)$)) × stpp(−$V(a,c)$,m) + (k_1 × (pow($V(a,c)$,3)) + k_2 × ($V(a,c)$)) × exp($I(G_1)$ × exp(−pow($V(Y)$/y_{on},2)) × $V(a,c)$/*sigmap*) × stpp($V(a,c)$,m))) × (1 − (pow((($V(Y)$ − stpp(−$V(a,c)$)),10)))}
*deriving the state variable as a voltage across the capacitor C_1 by G_2
C_1 *Y* 1 0 *IC* = 0.23
*a capacitor C_1 for obtaining the state variable
R_2 *Y* 0 10G
*additional resistor Rad for avoiding convergence issues
.func stpp(x,p) = {(1/2) × (1 + (x/sqrt(pow(x,2) + p)))}
*step-like differentiable function
ends *A10*

The described tantalum oxide memristor model is effectively analyzed by LTSPICE, version XVII, at various sinusoidal signals with various frequencies and amplitudes of the employed signal [12]. The obtained time diagrams and the *i–v* relationships are presented in Figure 4a–d to illustrate the proper operation of the described memristor model [12]. The state variable changes between 0 and 0.5. It starts from an initial value of 0.35 and, owing to the applied voltage and the generated flux linkage, their average value is decreased to a stable value of 0.25. This appropriate operation of the LTSPICE memristor model is supported by the detected decreasing of the area of the pinched current–voltage loop, while raising the frequency of the applied signal. Additional simulations at different amplitudes confirm the decrease of the area of the *i–v* loop. These are two of the main fingerprints of the memristor elements [7,12]. The SPICE programs simulate electronic schemes and devices using initialization and iterative procedures, numerically solving the corresponding differential equations by suitable solvers [23]. Convergence problems occur in the SPICE environment, mainly due to several reasons. Frequently, convergence issues occur in electronic circuits with several stable operating points. The limited precision of the numbers' representation is a reason for obtaining errors and convergence issues. During the numerical integration of the respective differential equations, some models are sensitive to the errors. For example, the capacitive element C_1, which is used as an integrator, is sensitive to the derived numerical errors and it accumulates them during the simulations. The truncation errors are also related to the simulations in SPICE. Such problems are also related to the rapid rise or fall of impulse currents and voltages and steep and discontinuous characteristics of electronic components. The convergence issues are related to erroneous results and sometimes to stopping the simulation before the previously defined end time, or to the impossibility of the simulation to start normally [23,33]. In some cases, the decrease of the time step or the change of the tolerances of some parameters could resolve these problems. In this sense, the use of a sigmoid step-like continuous function in modeling of memristors is another way to prevent the convergence problems in LTSPICE.

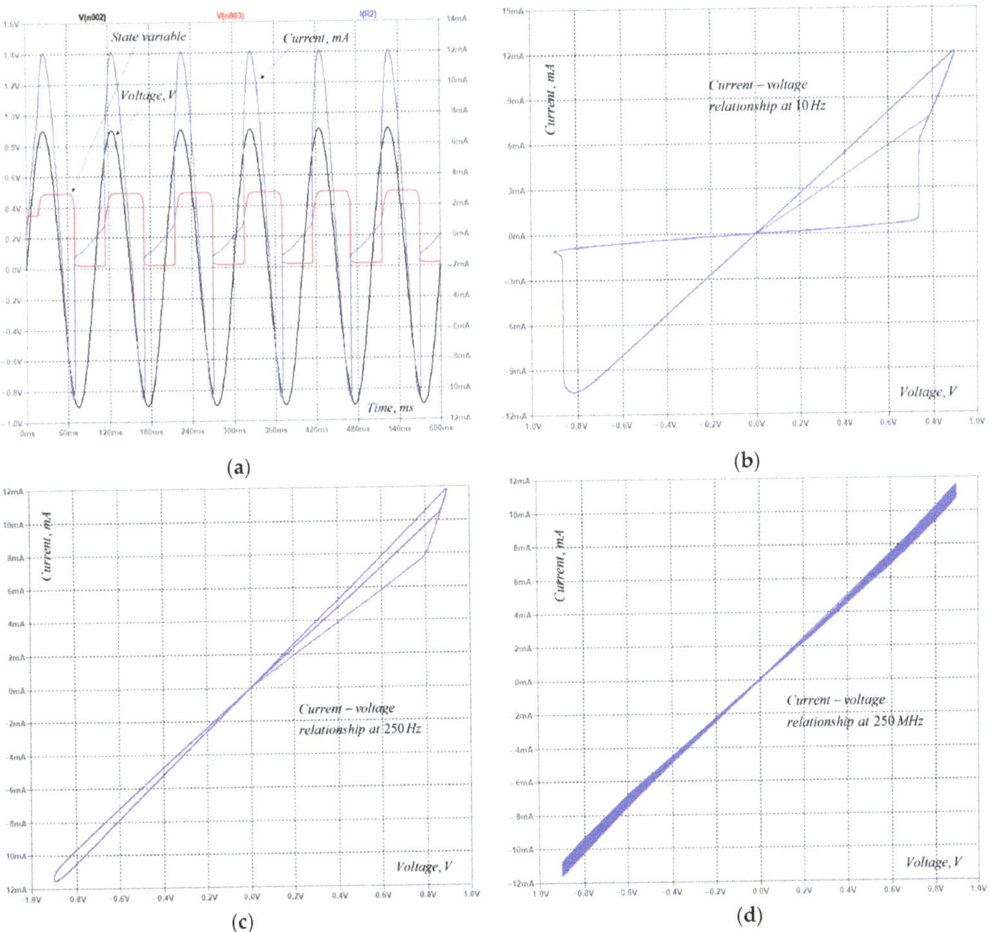

Figure 4. (**a**) Time graphs of memristor current and voltage with a frequency of 10 Hz. (**b**) Respective *i–v* relation of the modified model of memristor. (**c**) Voltage–current relation at a frequency of 250 Hz. (**d**) Current–voltage characteristic obtained for a frequency of 250 MHz.

The time step, used in the simulation setup, must be as low as possible, but it is related to the obtained output file and the simulation time of the respective scheme. According to the other SPICE products, LTSPICE is not sensitive to convergence issues. The computational efficiency of the considered memristor models is related to the amount of time and memory for a given iterative step in the simulation process. Conversely, the needed time for simulations is proportional to the number of elementary mathematical operations for obtaining the respective solution at a given iteration step [23,33].

The considered memristor model is also analyzed by rectangular pulse voltage signals with different duty cycles and amplitudes in the LTSPICE environment, and no convergence issues were observed [12].

5. Hybrid and Passive Memristor Memory Crossbars

In this section, the application of the considered modified tantalum oxide memristor model in hybrid and passive memory crossbars [12,29,30] is described, paying attention to the memristor operation in a hard-switching mode [7,8]. The main procedure for writing, reading, and erasing information is related to applying short and rectangular voltage pulses to the respective memristors and changing their resistance [12,29]. For reading the stored information without changing the corresponding content, voltage pulses with a level lower than the activation threshold are used [12,29,30].

5.1. A Simple Hybrid Memristor Crossbar

The considered model of the Ta_2O_5 memory element is tested and considered in a hybrid matrix [12,29,30]. A section of the memristor crossbar is presented in Figure 5a for better explanation of the basic writing, reading, and erasing procedures.

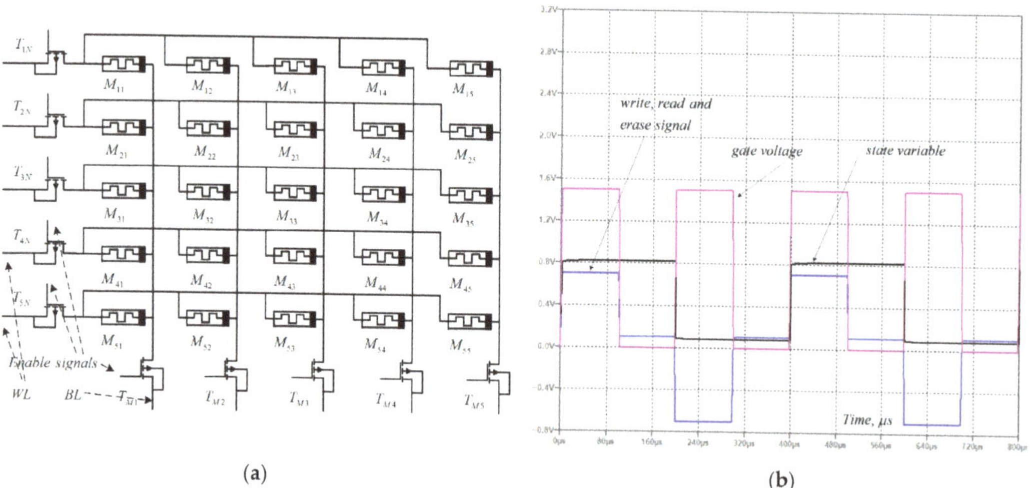

(a)

(b)

Figure 5. (a) A fragment of a simple hybrid memristor memory crossbar. (b) Time illustrations of the voltage across the memristor, the state variable x and the gate voltage.

Applying a positive rectangular voltage impulse with a duration of 100 µs and level of 0.7 V ensures the storing of a logical unity in the tantalum oxide memory component. The memristance of the element is changed to a minimal level, correspondent to ON-resistance status. When a voltage impulse with a negative polarity and identical duration and level is applied to the described memory element, it runs to its OFF-resistance state. Then, the correspondent memristance has a high-level value, and the logical data collected in the memristor is zero [12,30]. The technique for extracting the information accumulated in the respective memory element is accomplished by applying a voltage pulse with a duration of 100 µs and a level of 0.1 V. Due to the minimal level of the applied reading pulse, the memristor's status is not affected, and the correspondent data are not altered. The used MOS transistors are with N-channel, type Si4866DY. They are applied for realizing the read-allow, write-enable, and deleting of the written information in the memristor elements [12]. The applied gate voltage is a sequence of pulses with an amplitude of 1.5 V and a duration of 100 µs. It ensures the reading, writing, and erasing processes. When the gate voltage is zero, the respective memristor is isolated, and its state cannot be affected. The time graphs of the signal v and the state variable x of the element M_{11} are shown in Figure 5b. Their behavior validates the correct functioning of the memristor elements in a state close to a hard-switching mode.

5.2. Analysis of a Passive Memristor Crossbar

A simplified representation of a fragment of a passive memristor-based crossbar [12,29] is shown in Figure 6a for further explanation and discussion. This equivalent electric circuit is presented for further explanation of the memristor matrix structure and functioning of pulse and hard-switching mode. The memristor memory matrix is analyzed in the LTSPICE environment, applying the classical tantalum oxide memristor model [11], and its improved modification [12] is discussed in this work. The rectangular pulse voltage signals, used for writing, reading, and erasing logical information in the corresponding memristor cells, are applied between the respective word line and bit line of the selected memory element [12]. Due to the high resistance of the memristor elements in a reverse-biased state, the flowing sneak path currents do not significantly disturb the normal functioning of the considered memory crossbar. The respective time graphs of the applied voltage v, the state variable of the memristor x, and the corresponding current–voltage characteristic are illustrated in Figure 6b. These figures are presented to prove the proper memristor operation in a state near to a hard-switching mode [12]. The applied classical tantalum oxide memristor model [11], denoted by K_8 and its modified analogue and presented as A_{10} [26], have comparable behavior. For the writing, reading, and erasing processes in the memory matrix, different levels of voltage pulses are used. The amplitude of the voltage pulses for writing information is about 0.63 V. The duration of the applied pulses is about 200 µs. The respective alteration of the memristor state variable x is about 0.83. The correspondent change of the memristance is about 80% of the full resistance, ranging between R_{ON} and R_{OFF}. For reading the information accumulated in the respective memristors, positive pulses with lower amplitude (about 0.08 V) with the same duration are used. Due to the lower level of the reading voltage pulses, the state variable of the memristor element and the respective memristance does not alter; thus, the accumulated information is not altered.

For the process of erasing the information in the memristor cells, voltage pulses with negative polarity, a duration of 200 µs, and an amplitude of 0.63 V are applied. Due to the negative polarity impulses, the memristor state variable returns to its previous value. According to the alteration of the state variable in a broad range (approximately between 0.08 and 0.82), the memristor elements operate in a hard-switching mode.

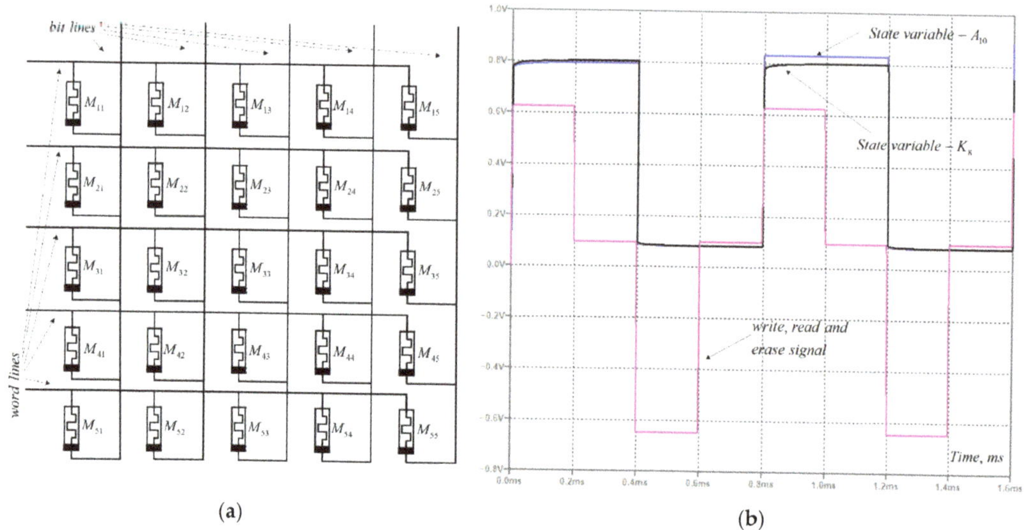

(a)   (b)

Figure 6. (**a**) A simple passive memristor memory crossbar. (**b**) Time diagrams of the memristor state variable according to models K_8 and A_{10} and the applied voltage.

6. Discussion

The modified and enhanced tantalum oxide memristor model considered in this paper is initially adjusted according to experimental current–voltage characteristics to derive the optimum values of the memory element model's factors. The optimization procedure is built on changing the model's coefficients and searching for the lowest RMS error among the current–voltage characteristics. The derived coefficients are used during the construction of the corresponding LTSPICE model. The generated LTSPICE memristor library model is analyzed and simulated for sinewave and impulse modes and for soft-switching and hard-switching representation. During the tests in an LTSPICE environment, no convergence issues are observed. The considered memristor model is investigated together with the classical and the existing tantalum oxide memristor models, and their identical behavior was confirmed. The detected reduction of the pinched current–voltage hysteresis loop's area with rising the operating frequency confirms the correctness of the considered memristor model. The state-flux relationships are also analyzed. For soft-switching mode, they are pinched hysteresis curves with a small area, and for hard-switching, they are hysteresis multi–valued curves with a broad area. For future work on oxide memristors' modeling and applications, additional attempts for simplification, generalization and optimizing the models will be conducted.

7. Conclusions

In the suggested modified and improved tantalum oxide memristor model, a simple altered window function built on the traditional Biolek's window is included. The offered model of the tantalum oxide memristor has a higher operation rate, regarding the classical and existing scientific literature modified models. It is tuned according to experimental current–voltage characteristics of Ta_2O_5 memristor nanostructures. It has a sufficient amount of coefficients for tuning and good correctness of the representation of the basic memristor's characteristics—current–voltage and state-flux relations. The described tantalum oxide memristor model has a relatively simple mathematical model and higher switching rate, according to the classical Hewlett–Packard model of the Ta_2O_5 memristor and its available modified versions. The used simplified window function successfully limits the memristor state variable x in the interval (0, 1), and it properly characterizes the boundary effects for a hard-switching operating state. The state terminal problems are resolved as well. An LTSPICE library memristor model is created in accordance with the considered model's mathematical equalities. It was suitably analyzed for sinusoidal and impulse modes, and convergence problems were not observed. The offered LTSPICE library memristor model was productively analyzed in hybrid and passive memory crossbars. The capability of the discussed tantalum oxide memory element model for suitable functioning in complex electrical circuits for both hard-switching and soft-switching and modes was proven.

Author Contributions: Conceptualization, V.M. and S.K.; methodology, V.M. and S.K.; software, V.M. and S.K.; validation, V.M. and S.K.; formal analysis, V.M.; investigation, V.M.; writing—original draft preparation, V.M. and S.K.; writing—review and editing, V.M.; visualization, V.M. and S.K.; supervision, V.M. All authors have read and agreed to the published version of the manuscript.

Funding: This research received no external funding.

Institutional Review Board Statement: Not applicable.

Informed Consent Statement: Not applicable.

Data Availability Statement: Not applicable.

Conflicts of Interest: The authors declare no conflict of interest.

References

1. Chiu, F.-C. A Review on Conduction Mechanisms in Dielectric Films. *Adv. Mater. Sci. Eng.* **2014**, *2014*, 1–18. [CrossRef]
2. Sawa, A. Resistive switching in transition metal oxides. *Mater. Today* **2008**, *11*, 28–36. [CrossRef]
3. Dearnaley, G.; Stoneham, A.M.; Morgan, D.V. Electrical phenomena in amorphous oxide films. *Rep. Prog. Phys.* **1970**, *33*, 1129–1191. [CrossRef]
4. Chiu, F.-C.; Li, P.-W.; Chang, W.-Y. Reliability characteristics and conduction mechanisms in resistive switching memory devices using ZnO thin films. *Nanoscale Res. Lett.* **2012**, *7*, 178. [CrossRef] [PubMed]
5. Chua, L. Memristor-The missing circuit element. *IEEE Trans. Circuit Theory* **1971**, *18*, 507–519. [CrossRef]
6. Strukov, D.B.; Snider, G.S.; Stewart, D.R.; Williams, R.S. The missing memristor found. *Nature* **2008**, *453*, 80–83. [CrossRef] [PubMed]
7. Biolek, Z.; Biolek, D.; Biolkova, V. Spice Model of Memristor with Nonlinear Dopant Drift. *Radioengineering* **2009**, *18*, 210–214.
8. Ascoli, A.; Tetzlaff, R.; Biolek, Z.; Kolka, Z.; Biolkova, V.; Biolek, D. The Art of Finding Accurate Memristor Model Solutions. *IEEE J. Emerg. Sel. Top. Circuits Syst.* **2015**, *5*, 133–142. [CrossRef]
9. Park, H.-L.; Kim, M.-H.; Lee, S.-H. Control of conductive filament growth in flexible organic memristor by polymer alignment. *Org. Electron.* **2020**, *87*, 105927. [CrossRef]
10. Yu, T.; He, F.; Zhao, J.; Zhou, Z.; Chang, J.; Chen, J.; Yan, X. Hf0.5Zr0.5O2-based ferroelectric memristor with multilevel storage potential and artificial synaptic plasticity. *Sci. China Mater.* **2021**, *64*, 727–738. [CrossRef]
11. Strachan, J.P.; Torrezan, A.C.; Miao, F.; Pickett, M.D.; Yang, J.J.; Yi, W.; Medeiros-Ribeiro, G.; Williams, S. State Dynamics and Modeling of Tantalum Oxide Memristors. *IEEE Trans. Electron Devices* **2013**, *60*, 2194–2202. [CrossRef]
12. Mladenov, V.; Kirilov, S. A Simplified Model of Tantalum Oxide Based Memristor and Application in Memory Crossbars. In Proceedings of the 2021 10th International Conference on Modern Circuits and Systems Technologies (MOCAST), Thessaloniki, Greece, 5–7 July 2021; pp. 1–4. [CrossRef]
13. Ascoli, A.; Tetzlaff, R.; Chua, L. Robust Simulation of a TaO Memristor Model. *Radioengineering* **2015**, *24*, 384–392. [CrossRef]
14. Ntinas, V.; Ascoli, A.; Tetzlaff, R.; Sirakoulis, G.C. Transformation techniques applied to a TaO memristor model to enable stable device simulations. In Proceedings of the 2017 European Conference on Circuit Theory and Design (ECCTD), Catania, Italy, 4–6 September 2017; pp. 1–4. [CrossRef]
15. Mladenov, V. A Modified Tantalum Oxide Memristor Model for Neural Networks with Memristor-Based Synapses. In Proceedings of the 2020 9th International Conference on Modern Circuits and Systems Technologies (MOCAST), Bremen, Germany, 7–9 September 2020; pp. 1–4. [CrossRef]
16. Parit, A.K.; Yadav, M.S.; Gupta, A.K.; Mikhaylov, A.; Rawat, B. Design and modeling of niobium oxide-tantalum oxide based self-selective memristor for large-scale crossbar memory. *Chaos Solitons Fractals* **2021**, *145*, 110818. [CrossRef]
17. Jin, S.; Kwon, J.-D.; Kim, Y. Statistical Analysis of Uniform Switching Characteristics of Ta2O5-Based Memristors by Embedding In-Situ Grown 2D-MoS2 Buffer Layers. *Materials* **2021**, *14*, 6275. [CrossRef] [PubMed]
18. Ryu, J.-H.; Hussain, F.; Mahata, C.; Ismail, M.; Abbas, Y.; Kim, M.-H.; Choi, C.; Park, B.-G.; Kim, S. Filamentary and interface switching of CMOS-compatible Ta2O5 memristor for non-volatile memory and synaptic devices. *Appl. Surf. Sci.* **2020**, *529*, 147167. [CrossRef]

19. Miao, F.; Yi, W.; Goldfarb, I.; Yang, J.J.; Zhang, M.-X.; Pickett, M.D.; Strachan, J.P.; Ribeiro, G.M.; Williams, S. Continuous Electrical Tuning of the Chemical Composition of TaOx-Based Memristors. *ACS Nano* **2012**, *6*, 2312–2318. [CrossRef]
20. Ryu, J.-H.; Mahata, C.; Kim, S. Long-term and short-term plasticity of Ta2O5/HfO2 memristor for hardware neuromorphic application. *J. Alloy. Compd.* **2021**, *850*, 156675. [CrossRef]
21. Tian, W.; Ilyas, N.; Li, D.; Li, C.; Jiang, X.; Li, W. Reliable Resistive Switching Behaviour of Ag/Ta2O5/Al2O3/p++-Si Memory Device. *J. Phys. Conf. Ser.* **2020**, *1637*, 012021. [CrossRef]
22. Yang, Y.; Lee, S.C. *Circuit Systems with MATLAB and PSpice*; John Wiley & Sons: Hoboken, NJ, USA, 2008; ISBN 978-04-7082-240-1.
23. May, C. *Passive Circuit Analysis with LTspice®: An Interactive Approach*; Springer Nature: Cham, Switzerland, 2020; p. 763. ISBN 978-3-030-38304-6. [CrossRef]
24. Chen, S.; Billings, S.A.; Luo, W. Orthogonal least squares methods and their application to non-linear system identification. *Int. J. Control* **1989**, *50*, 1873–1896. [CrossRef]
25. Iliev, A.; Kyurkchiev, N.; Markov, S. On the approximation of the step function by some sigmoid functions. *Math. Comput. Simul.* **2017**, *133*, 223–234. [CrossRef]
26. Mladenov, V. A Unified and Open LTSPICE Memristor Model Library. *Electronics* **2021**, *10*, 1594. [CrossRef]
27. Dozortsev, A.; Goldshtein, I.; Kvatinsky, S. Analysis of the row grounding technique in a memristor-based crossbar array. *Int. J. Circuit Theory Appl.* **2018**, *46*, 122–137. [CrossRef]
28. Mladenov, V. *Advanced Memristor Modeling: Memristor Circuits and Networks*; MDPI: Basel, Switzerland, 2019; p. 170. [CrossRef]
29. Yakopcic, C.; Taha, T.M.; Hasan, R. Hybrid crossbar architecture for a memristor based memory. In Proceedings of the NAECON 2014—IEEE National Aerospace and Electronics Conference, Dayton, OH, USA, 24–27 June 2014; pp. 237–242. [CrossRef]
30. Qureshi, M.S.; Pickett, M.; Miao, F.; Strachan, J.P. CMOS interface circuits for reading and writing memristor crossbar array. In Proceedings of the 2011 IEEE International Symposium of Circuits and Systems (ISCAS), Rio de Janeiro, Brazil, 19–18 May 2011; pp. 2954–2957. [CrossRef]
31. Ascoli, A.; Corinto, F.; Senger, V.; Tetzlaff, R. Memristor Model Comparison. *IEEE Circuits Syst. Mag.* **2013**, *13*, 89–105. [CrossRef]
32. Mandic, D.P. A Generalized Normalized Gradient Descent Algorithm. *IEEE Signal Process. Lett.* **2004**, *11*, 115–118. [CrossRef]
33. Biolek, D.; Di Ventra, M.; Pershin, Y.V. Reliable SPICE simulations of memristors, memcapacitors and meminductors. *arXiv* **2013**, arXiv:1307.2717.

 *technologies*

Article

Artwork Style Recognition Using Vision Transformers and MLP Mixer

Lazaros Alexios Iliadis [1], Spyridon Nikolaidis [1], Panagiotis Sarigiannidis [2] and Shaohua Wan [3] and Sotirios K. Goudos [1,*]

1. ELEDIA@AUTH, School of Physics, Aristotle University of Thessaloniki, 541 24 Thessaloniki, Greece; liliadis@physics.auth.gr (L.A.I.); snikolaid@physics.auth.gr (S.N.)
2. Department of Informatics and Telecommunications Engineering, University of Western Macedonia, 501 00 Kozani, Greece; psarigiannidis@uowm.gr
3. School of Information and Safety Engineering, Zhongnan University of Economics and Law, Wuhan 430073, China; shwanhust@zuel.edu.cn
* Correspondence: sgoudo@physics.auth.gr

Abstract: Through the extensive study of transformers, attention mechanisms have emerged as potentially more powerful than sequential recurrent processing and convolution. In this realm, Vision Transformers have gained much research interest, since their architecture changes the dominant paradigm in Computer Vision. An interesting and difficult task in this field is the classification of artwork styles, since the artistic style of a painting is a descriptor that captures rich information about the painting. In this paper, two different Deep Learning architectures—Vision Transformer and MLP Mixer (Multi-layer Perceptron Mixer)—are trained from scratch in the task of artwork style recognition, achieving over 39% prediction accuracy for 21 style classes on the WikiArt paintings dataset. In addition, a comparative study between the most common optimizers was conducted obtaining useful information for future studies.

Keywords: vision transformers; computer vision; deep learning; artistic style recognition

Citation: Iliadis, L.A.; Nikolaidis, S.; Sarigiannidis, P.; Wan, S.; Goudos, S.K. Artwork Style Recognition Using Vision Transformers and MLP Mixer. *Technologies* **2022**, *10*, 2. https://doi.org/10.3390/technologies10010002

Academic Editor: Pietro Zanuttigh

Received: 18 November 2021
Accepted: 23 December 2021
Published: 28 December 2021

Publisher's Note: MDPI stays neutral with regard to jurisdictional claims in published maps and institutional affiliations.

1. Introduction

Deep Learning (DL) has become the dominant paradigm in the field of Computer Vision (CV). ImageNet Challenge (ILSVRC) has been proven to be a driving force for many novel neural network architectures, which have prevailed in the CV research community [1]. However, there are many CV tasks that include the classification of more abstract visual forms, like clouds in the sky [2], abstract images [3] and paintings [4].

The artistic style (or artistic movement) of a painting is a descriptor that contains valuable data about the painting itself, providing, at the same time, a framework of reference for further analysis. In this context, artistic style recognition is an important task in CV taking into consideration that authentic artwork carries a high value (aesthetic, historic and economic) [4]. Artwork style recognition, artist classification and other CV tasks related to paintings had been studied before the "DL revolution" [5–7].

A great deal of research work has been done in this field leading to many impressive results [4–12]. Although many techniques have been deployed, the architectures based on Convolutional Neural Networks (CNNs) have prevailed. Nevertheless, the current technology appears to have reached a plateau in model performance, highlighting the need for new designs. In recent months, there have been a plethora of new DL-based CV models that perform really well on many tasks. However, these models have not been tested, to the best of our knowledge, on the artwork style recognition task.

Transformers have become the dominant architecture in use in the field of Natural Language Processing (NLP), outperforming previous models in various tasks [13]. Transformers are based on the attention mechanism. Attention allows to derive information from any state of a given text sequence. Introducing the attention layer, it is possible to access all previous states and weigh them according to a learned measure of relevancy to the current token, providing sharper information about far-away relevant tokens [13]. Transformers have proven that the recurrence is unnecessary.

In addition, modified architectures have been successfully applied on object detection tasks [14]. Until recently in the task of image recognition, attention has been complementary to convolutions. Extending the idea of attention only mechanisms to CV, a few modifications to the basic transformer architecture are required [15].

Another recent proposal in DL-based CV is MLP Mixer [16]. MLP Mixer is based solely on multi-layer perceptrons (MLPs) and does not make use of either convolutions or attention mechanisms. MLP Mixer may be proven to be a valid alternative to many CV tasks, where there are not so many training data or where the available hardware does not support more expensive (in computational terms) architectures.

Following [17], we present the motivation for this work, along with the letter contributions and the organization of the rest of this work.

1.1. Motivation

The guiding motivation for this research is twofold. On the one hand, there is a need to test the newly proposed DL models on more complicated CV tasks and to demonstrate their applicability. On the other hand, artwork style recognition is a complex problem that needs to be studied further from the research community, since it poses interesting questions about aesthetics, artistic movements, the connection between different styles etc.

1.2. Contribution

The main contributions of this work are as follows

- We propose Vision Transformers as the main ML method to classify artistic style.
- We train Vision Transformers from scratch in the task of artwork style recognition, achieving over 39% prediction accuracy for 21 style classes on the WikiArt paintings dataset.
- We conduct a comparative study between the most common optimizers obtaining useful information for future studies.
- We compare the results compared with MLP Mixer's performance on the same task, examining in this way two very different DL architectures on a complex pattern recognition framework.

To the best of the authors knowledge, this is the first time that Vision Transformers have been applied to the specific problem. The results obtained in this work provide a minimum benchmark for future studies regarding the application of ViT and MLP Mixer in the artwork style recognition task and possibly to other CV tasks, which may include a diverse set of training images.

1.3. Organization of the Paper

The rest of the paper is organized as follows: Section 2 briefly describes related work. In Section 3 Transformers, Vision transformers, MLP Mixer and the basic information about DL optimizers are discussed and details about WikiArt paintings dataset are provided. We elaborate and present the numerical results in Section 4. Finally, Section 5 concludes this work.

2. Related Work

ML and DL techniques have been successfully deployed in the task of Artistic Style recognition. In [4], researchers conducted a comprehensive study of CNNs applied to the task of style classification of paintings and analyzed the learned representation through correlation analysis with concepts derived from art history. In [8–12], many DL and Image Processing techniques are deployed in order to improve accuracy. The advantages and disadvantages of these methods are presented in Table 1 following the presentation in [18].

Another field of study is the Image Style Transfer. Gatys et al. [19], by separating and recombining the image content and style of images, managed to produce new images that combine the content of an arbitrary photograph with the appearance of numerous well-known artworks. Many modifications and optimizations have been proposed since [20,21]. However, in style transfer, it is necessary to separate style from content as much as possible, whereas, in artistic style recognition, the description of the content is used as an additional feature [8].

An active research area is the use of a Generative Adversarial Network (GAN) for conditional image synthesis (ArtGAN) [22,23]. The proposed model is capable of creating realistic artwork as well as generate compelling real world images.

Vision Transformers have gained much research interest. The first model based solely on attention is ViT [15], while [16] introduces MLP Mixer. To the best of our knowledge, this is the first time that ViT and MLP Mixer are implemented on the task of artistic style classification.

Table 1. Artwork style recognition based on DL methods.

Paper	Advantages	Disadvantages
Elgammal A., et al. [4]	• Study of many CNN architectures • Interpretation and representation	No comparison with previous works
Lecoutre A., et. al. [8]	• Comprehensive methodology • Plenty techniques used	Full analysis is provided only for Alexnet
Bar Y., et. al. [21]	• Combination of low level descriptors and CNNs	test only one CNN architecture
Cetinic E., et. al. [10]	• Fine-tuning • Analysing image similarity	No interpretation
Huang X., et. al. [11]	• Two channels used; the RGB channel and the brush stroke information	No interpretation
Sandoval C., et al. [12]	• Novel two stage approach	Only pre-trained models

3. Materials and Methods

3.1. Vision Transformers

Vision Transformer (ViT) was proposed as an alternative to convolutions in deep neural networks. The model was pre-trained on a large dataset of images collected by Google and later fine-tuned to downstream recognition benchmarks. A large dataset is necessary in order to achieve state of the art results.

The main architecture of the model is depicted in Figures 1 and 2. ViT processes 2D images patches that are flattened in a vector form and fed to the transformer as a sequence. These vectorized patches are then projected to a patch embedding using a linear layer, and position embedding is attached to encode location information. In addition, at the beginning of the input, a classification token is attached to the transformer. The output representation corresponding to the first position is then used as the global image representation for the image classification task.

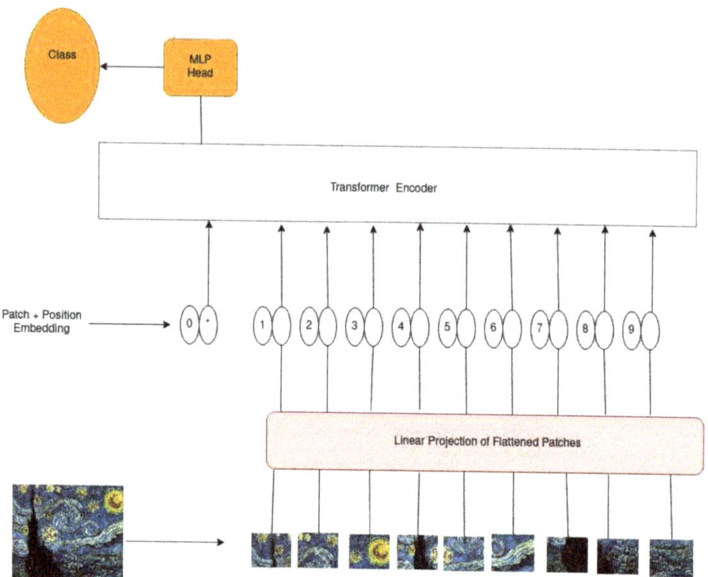

Figure 1. ViT model overview.

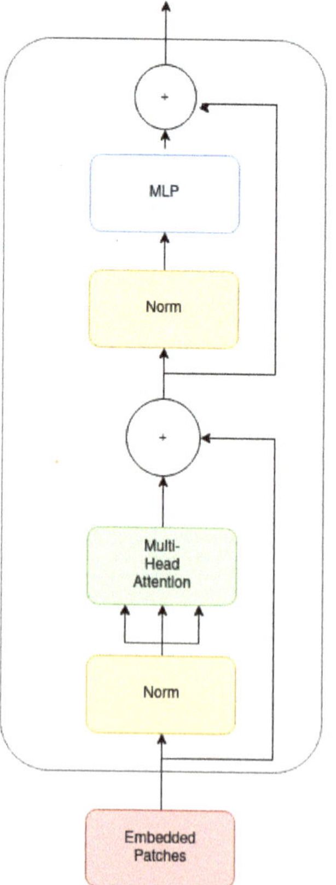

Figure 2. Transformer encoder: The illustration is inspired by [15].

3.2. MLP Mixer

MLP Mixer was recently introduced [16] as a CV model based solely on MLPs, without using either convolutions or attention mechanisms. The core idea behind MLP Mixer is to separate in a clear way using only MLPs, the per-location (channel-mixing) operations and the cross-location (token-mixing) operations.

Mixer takes as input a sequence of S non-overlapping image patches, each one projected to a desired hidden dimension C, thus obtaining a matrix $X \in \mathbb{R}^{S \times C}$. If the original input image has resolution (H, W), and each patch has resolution (P, P), then the number of patches is $S = \frac{H \cdot W}{P^2}$. All patches are linearly projected with the same projection matrix. Mixer consists of multiple layers of identical size, and each layer consists of two MLP blocks.

- The token-mixing MLP: it acts on columns of X and is shared across all columns.
- The channel-mixing MLP: it acts on rows of X and is shared across all rows.

Unlike ViT, MLP Mixer does not use position embeddings because the token-mixing MLPs are sensitive to the order of the input tokens.

3.3. Optimizers

Hyper-parameter optimization is a crucial part of DL training process. Image classification is usually considered a Supervised Learning task. In this framework, given a dataset, the learning algorithm is trained in such a way to minimize a suitably chosen cost function $\mathcal{L}$. An optimizer is needed to achieve the minimum of this function. For an extensive review of the most common used optimizers in DL, one may refer to [24]. Here the weights' update rule for each method is provided. In order to evaluate each optimizer "equally", none of the "tricks" that are proposed in the literature were used.

In the following, θ_t means the weight at step t, η_t is the learning rate, and v_t is the momentum. The rest of the parameters are explained in [24–26].

1. Stochastic Gradient Descent: Stochastic Gradient Descent (SGD) is one of the most used optimizers. SGD allows to update the network weights per each training image (online training).

$$\theta_{t+1} = \theta_t - \eta_t \nabla \mathcal{L}(\theta_t) \tag{1}$$

2. Momentum Gradient Descent: SGD may lead to oscillations during training. The best way to avoid them is the knowledge of the right direction for the gradient. This information is derived from the previous position, and, when considering the previous position, the updating rule adds a fraction of the previous update, which gives the optimizer the momentum needed to continue moving in the right direction. The weights in the Momentum Gradient Descent (MGD) are updated as

$$v_0 = 0, \quad v_{t+1} = \gamma v_t + \nabla \mathcal{L}(\theta_t)$$
$$\theta_{t+1} = \theta_t - \eta_t \nabla \mathcal{L}(\theta_t) \tag{2}$$

3. Adam: Adam has been introduced as an algorithm for the first-order gradient-based optimization of stochastic objective functions, based on adaptive estimates of lower-order moments [25]. Adam has been established as one of the most successful optimizers in DL.

$$m_0 = 0, \quad v_0 = 0$$
$$m_{t+1} = \beta_1 m_t + (1 - \beta_1) \nabla \mathcal{L}(\theta_t)$$
$$v_{t+1} = \beta_2 v_t + (1 - \beta_2) \nabla \mathcal{L}(\theta_t)^2$$
$$b_{t+1} = \frac{\sqrt{1 - \beta_2^{t+1}}}{1 - \beta_1^{t+1}} \tag{3}$$
$$\theta_{t+1} = \theta_t - a_t \frac{m_{t+1}}{\sqrt{v_{t+1} + \epsilon}} b_{t+1}$$

4. AdaMax: AdaMax is a generalisation of Adam from the l_2 norm to the l_∞ norm [25].
5. Optimistic Adam: Optimistic Adam (OAdam) optimizer [26] is a variant of the ADAM optimizer. The only difference between OAdam and Adam is the weight update,

$$\theta_{t+1} = \theta_t - 2\frac{\eta_t}{\sqrt{v_{t+1}+\epsilon}}b_{t+1} + \frac{\eta_t}{\sqrt{v_t+\epsilon}}b_t \tag{4}$$

6. RMSProp: Using some Adaptive Gradient Descent Optimizers leads, in some cases, the learning rate to decrease monotonically because every added term is positive. After many epochs, the learning rate is so small that it stops updating the weights. The RMSProp method proposes

$$\begin{aligned} v_0 &= 1, \quad m_0 = 0 \\ v_{t+1} &= \rho v_t + (1-\rho)\nabla\mathcal{L}(\theta_t)^2 \\ m_{t+1} &= \gamma m_t + \frac{\eta_t}{\sqrt{v_{t+1}+\epsilon}} \\ \theta_{t+1} &= \theta_t - m_{t+1} \end{aligned} \tag{5}$$

3.4. WikiArt Dataset

The dataset used in this work is called WikiArt paintings dataset and has been collected from the ArtGan [22] GitHub repository website (https://github.com/cs-chan/ArtGAN/tree/master/WikiArt%20Dataset, accessed on 20 July 2021) with its respective metadata.

The dataset contains 81,446 images tagged with one corresponding style among the 27 following styles: Abstract Expressionism (2782 images), Action Painting (92 images), Analytical Cubism (110 images), Art Nouveau (4334 images), Baroque (4241 images), Color Field Painting (1615 images), Contemporary Realism (481 images), Cubism (2235 images), Early Renaissance (1391 images), Expressionism (6736 images), Fauvism (934 images), High Renaissance (1343 images), Impressionism (13,060), Mannerism (Late Renaissance) (1279 images), Minimalism (1337 images), Naive Art/Primitivism (2405 images), New Realism (314 images), Northern Renaissance (2552 images), Pointillism (513 images), Pop Art (1483 images), Post Impressionism (6451 images), Realism (10,733 images), Rococo (2089 images), Romanticism (7049 images), Symbolism (4528 images), Synthetic Cubism (216 images) and Ukiyo-e (1167 images).

The WikiArt dataset is highly unbalanced. To avoid some of the issues that may follow, Action Painting and Pointillism classes were dropped. In addition, Analytical Cubism and Synthetic Cubism classes were incorporated into the Cubism class and similarly Contemporary Realism and New Realism were transferred into Realism class. The resulting dataset was comprised of 21 classes and 80,835 images in total. The train, validation and test sets were 60%, 20% and 20% of the whole dataset respectively.

In Figure 3, two samples of the dataset are shown, highlighting the diversity of the artwork style.

(a)   (b)

Figure 3. Samples from the WikiArt dataset. (**a**) Mona Lisa; (**b**) Composition VIII.

4. Results and Discussion

4.1. Experiments

Two types of experiments were conducted: In the first, a comparative study between the most common used optimizers in DL was conducted as part of ViT training process. In the second experiment, MLP Mixer was trained from scratch on the same task, drawing in this way useful information for future studies regarding complex CV problems. The dataset was pre-processed as described in the previous section.

4.1.1. ViT: Optimizers' Performance

For the purposes of this study, ViT is comprised of eight heads, the dropout is set to 0.2, and the image size is 256×256. The training was performed on a single NVIDIA RTX 2070 GPU, for 100 epochs. The experiments were conducted using Python DL libraries Pytorch and Sci-kit learn, along with the "classical" ones like Matplotlib, Pandas and Numpy.

Since many optimizers are frequently used in different CV tasks, in the first experiment a comparative study between the most common used ones, was conducted. The scope of this experiment is to evaluate ViT's performance on the task of artistic style recognition, setting in the same time a minimum benchmark for future studies. Five-fold cross-validation was used. The results are shown in Table 2.

Table 2. ViT performance in artwork style recognition.

Optimizer	Accuracy
Adam	39.89%
Adamax	39.42%
Optimistic Adam	39.71%
SGD	39.28%
MGD	39.31%
RMSProp	38.97%

From Table 2 it follows that the Adam optimizer outperforms the others, achieving over 39% prediction accuracy. The test loss for the best optimizer is depicted in Figure 4. The loss is calculated on test set and it shows how well the model is doing for this set. The loss function that is used is negative log-likelihood. It is clear that that the test procedure was stable enough, and it is depicted that the model learns through training. Table 3 provides a better visualization of the results since it shows the prediction accuracy per class. However, it should be noted that the difference in the results is not significant between the different optimizers.

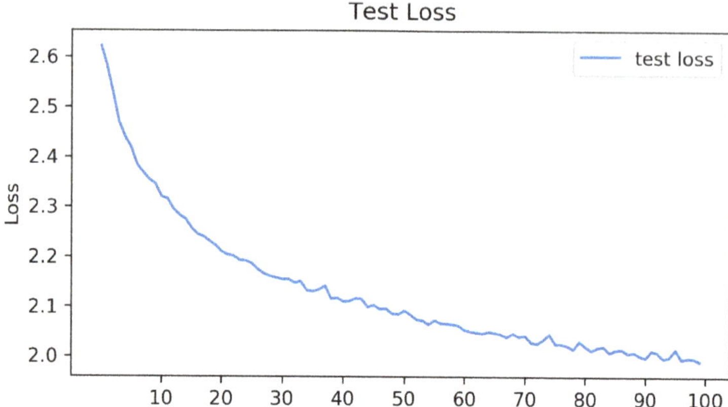

Figure 4. ViT: Adam training test loss.

Table 3. ViT accuracy per class.

Class	Accuracy %
Abstract Expressionism	29.6
Art Nouveau	25.3
Baroque	48.3
Color Field Painting	65.5
Cubism	21.3
Early Renaissance	34.1
Expressionism	28.0
Fauvism	17.5
High Renaissance	8.5
Impressionism	65.0
Mannerism Late Renaissance	18.1
Minimalism	49.4
Naive Art / Primitivism	15.4
Northern Renaissance	6.1
Pop Art	14.6
Post Impressionism	30.6
Realism	57.7
Rococo	45.3
Romanticism	37.4
Symbolism	28.4
Ukiyo-e	68.3

Table 3 shows that the model performs really well on the Ukiyo-e class, the Color Field Painting class, Impressionism and the Realism class and also achieves a good performance in Baroque and Rococo.

4.1.2. MLP Mixer Performance

In the second experiment MLP Mixer, a recent DL architecture that does not use either convolutions or attention, was trained from scratch on artwork style recognition. The scope of this experiment is to compare the two advancements in CV architectures and to determine if a lighter model, like MLP Mixer, is able to achieve good results on a complex CV task.

For the purposes of this study, MLP Mixer has a depth equal to 8, the dropout is set to 0.3, and the image size is 256×256. The training was also performed on a single

NVIDIA RTX 2070 GPU, for 100 epochs. The experiments were conducted using Python DL libraries Pytorch and Sci-kit learn, along with the "classical" ones like Matplotlib, Pandas and Numpy.

The test loss in the test set is shown in Figure 5. The results show that the learning process was not stable. Perhaps more work on the estimation of model's hyper-parameters is needed. However, the MLP Mixer achieves a prediction accuracy similar to that of ViT Table 4. This result indicates that a lighter architecture can be applied in such complicated tasks. For a better visualization of the results, a matrix showing the accuracy per class is built Table 5.

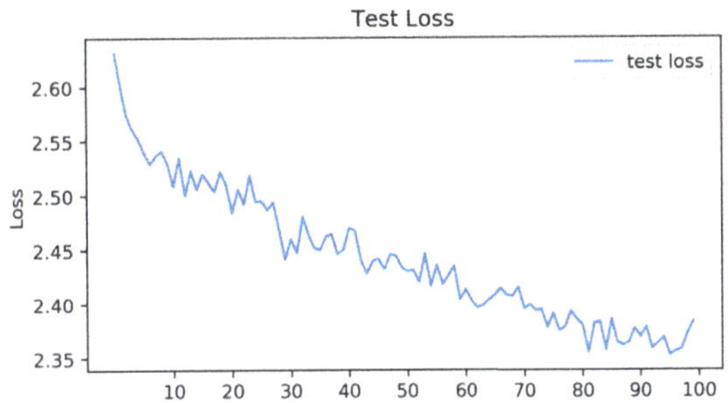

Figure 5. MLP Mixer: test loss.

Table 4. MLP Mixer performance in artwork style recognition.

Model	Accuracy
MLP Mixer	39.59%

Table 5. MLP Mixer accuracy per class.

Class	Accuracy %
Abstract Expressionism	29.0
Art Nouveau	34.3
Baroque	46.2
Color Field Painting	64.9
Cubism	20.4
Early Renaissance	34.1
Expressionism	28.0
Fauvism	16.8
High Renaissance	23.1
Impressionism	62.0
Mannerism Late Renaissance	15.1
Minimalism	46.6
Naive Art / Primitivism	25.5
Northern Renaissance	5.3

Table 5. *Cont*

Class	Accuracy %
Pop Art	22.1
Post Impressionism	30.4
Realism	44.3
Rococo	44.4
Romanticism	36.2
Symbolism	28.2
Ukiyo-e	60.9

Table 5 shows that the model performed well on the Ukiyo-e class, Color Field Painting class and Impressionism class and also achieved good performance in Baroque, Realism and Rococo. ViT and MLP Mixer learned different classes better, and a further investigation of the learned parameters is set as a future goal.

5. Conclusions

In this paper, the Vision Transformers ViT and MLP Mixer were successfully applied on the WikiArt dataset in the artistic style recognition task. ViT was trained from scratch in the WikiArt dataset achieving over 39% accuracy for 21 classes, thus, setting a minimum benchmark in accuracy prediction for future studies. In addition, a comparative study was conducted among the most common used optimizers, which showed that training with the Adam optimizer and Optimistic Adam optimizer resulted in better performance. Using the above results, MLP Mixer was trained from scratch, performing close to ViT in terms of prediction accuracy. As suggested by our experiments and literature, the use of larger datasets with richer resources should improve the accuracy of the models.

Future work on this subject will be focused on improvements on the models' hyperparameters through parametric studies and other experiments. Variations of the models that were used here may provide better results, especially with the combination of other CV techniques. In addition, the creation of a larger dataset will provide a better overview of the tested models' prediction accuracy.

Author Contributions: Conceptualization, S.K.G. and L.A.I.; methodology, L.A.I.; software, L.A.I.; validation, S.N., P.S., S.W. and S.K.G; formal analysis, L.A.I. and S.K.G.; investigation, S.N., P.S. and S.W.; resources, S.N.; data curation, L.A.I.; writing—original draft preparation, L.A.I.; writing—review and editing, S.N., P.S., S.W. and S.K.G.; visualization, L.A.I.; supervision, S.K.G.; project administration, S.K.G. All authors have read and agreed to the published version of the manuscript.

Funding: This research received no external funding.

Institutional Review Board Statement: Not applicable.

Informed Consent Statement: Not applicable.

Data Availability Statement: WikiArt dataset was downlowded from the following Github repository https://github.com/cs-chan/ArtGAN/tree/master/WikiArt%20Dataset and it was accessed on 20 July 2021.

Conflicts of Interest: The authors declare no conflict of interest.

Abbreviations

The following abbreviations are used in this manuscript:

CNN   Convolutional Neural Networks
CV   Computer Vision
DL   Deep Learning
GAN   Generative Adversarial Network
MLP   Multi-Layered Perceptron
NLP   Natural Language Processing
ViT   Visual Transformer

References

1. Waseem, R.; Zenghui, W. Deep Convolutional Neural Networks for Image Classification: A Comprehensive Review. *Neural Comput.* **2017**, *29*, 2352–2449. [CrossRef]
2. Olejnik, A.; Borecki, M.; Rychlik, A. A simple detection method of movement of clouds at the sky. In Proceedings of the SPIE 11581, Photonics Applications in Astronomy, Communications, Industry, and High Energy Physics Experiments, Wilga, Poland, 14 October 2020; p. 1158111. [CrossRef]
3. Stabinger, S.; Rodríguez-Sánchez, A. Evaluation of Deep Learning on an Abstract Image Classification Dataset. In Proceedings of the 2017 IEEE International Conference on Computer Vision Workshops (ICCVW), Venice, Italy, 22–29 October 2017; pp. 2767–2772. [CrossRef]
4. Elgammal, A.; Liu, B.; Kim, D.; Elhoseiny, M. The Shape of Art History in the Eyes of the Machine. In Proceedings of the AAAI, Palo Alto, CA, USA, 2–7 February 2018.
5. Johnson, C.R.; Hendriks, E.; Berezhnoy, I.J.; Brevdo, E.; Hughes, S.M.; Daubechies, I.; Li, J.; Postma, E.; Wang, J.Z. Image processing for artist identification. *IEEE Signal Process. Mag.* **2008**, *25*, 37–48. [CrossRef]
6. Altenburgera, P.; Kämpferb, P.; Makristathisc, A.; Lubitza, W.; Bussea, H.-J. Classification of bacteria isolated from a medieval wall painting. *J. Biotechnol.* **1996**, *47*, 39–52. [CrossRef]
7. Li, C.; Chen, T. Aesthetic Visual Quality Assessment of Paintings. *IEEE J. Sel. Top. Signal Process.* **2009**, *3*, 236–252. [CrossRef]
8. Lecoutre, A.; Negrevergne, B.; Yger, F. Recognizing Art Style Automatically in Painting with Deep Learning. In Proceedings of the Ninth Asian Conference on Machine Learning, Seoul, Korea, 15–17 November 2017; pp. 327–342.
9. Bar, Y.; Levy, N.; Wolf, L. Classification of Artistic Styles Using Binarized Features Derived from a Deep Neural Network. In *Lecture Notes in Computer Science Proceedings of the ECCV Workshops, Zurich, Switzerland, 6–7 September 2014*; Springer: Cham, Switzerland, 2014.
10. Cetinic, E.; Lipic, T.; Grgic, S. Fine-tuning Convolutional Neural Networks for Fine Art Classification. *Expert Syst. Appl.* **2018**, *114*, 107–118. [CrossRef]
11. Huang, X.; Zhong, S.; Zhijiao, X. Fine-Art Painting Classification via Two-Channel Deep Residual Network. In *Lecture Notes in Computer Science, Proceedings of the Advances in Multimedia Information Processing, Harbin, China, 28–29 September 2017*; Springer: Cham, Switzerland, 2017. [CrossRef]
12. Sandoval, C.; Pirogova, E.; Lech, M. Two-Stage Deep Learning Approach to the Classification of Fine-Art Paintings. *IEEE Access* **2019**, *7*, 41770–41781. [CrossRef]
13. Vaswani, A.; Shazeer, N.; Parmar, N.; Uszkoreit, J.; Jones, L.; Gomez, A.; Kaiser, L.; Polosukhin, I. Attention is all you need. In Proceedings of the 31st International Conference on Neural Information Processing Systems (NIPS'17), Long Beach, CA, USA, 4–9 December 2017; Curran Associates Inc.: Red Hook, NY, USA, 2017; pp. 6000–6010.
14. Carion, N.; Massa, F.; Synnaeve, G.; Usunier, N.; Kirillov, A.; Zagoruyko, S. End-to-End Object Detection with Transformers. In *Lecture Notes in Computer Science, Proceedings of the Computer Vision – ECCV 2020, Glasgow, UK, 23–28 August 2020*; Vedaldi, A., Bischof, H., Brox, T., Frahm, J.M., Eds.; Springer: Cham, Switzerland, 2020; Volume 12346. [CrossRef]
15. Dosovitskiy, A.; Beyer, L.; Kolesnikov, A.; Weissenborn, D.; Zhai, X.; Unterthiner, T.; Gelly, S. An Image is Worth 16×16 Words: Transformers for Image Recognition at Scale. In Proceedings of the ICLR 2021: The Ninth International Conference on Learning Representations, Vienna, Austria, 3–7 May 2021.
16. Tolstikhin, I.O.; Houlsby, N.; Kolesnikov, A.; Beyer, L.; Zhai, X.; Unterthiner, T.; Yung, J.; Keysers, D.; Uszkoreit, J.; Lucic, M.; et al. MLP-Mixer: An all-MLP Architecture for Vision. *arXiv* **2021**, arXiv:2105.01601.
17. Amin, F.; Choi, G.S. Advanced Service Search Model for Higher Network Navigation Using Small World Networks. *IEEE Access* **2021**, *9*, 70584–70595. [CrossRef]
18. Amin, F.; Ahmad, A.; Sang Choi, G. Towards Trust and Friendliness Approaches in the Social Internet of Things. *Appl. Sci.* **2019**, *9*, 166. [CrossRef]
19. Gatys, L.; Ecker, A.; Bethge, M. Image Style Transfer Using Convolutional Neural Networks. In Proceedings of the 2016 IEEE Conference on Computer Vision and Pattern Recognition (CVPR), Las Vegas, NV, USA, 26 June–1 July 2016; pp. 2414–2423. [CrossRef]
20. Gatys, L.A.; Ecker, A.S.; Bethge, M.; Hertzmann, A.; Shechtman, E. Controlling Perceptual Factors in Neural Style Transfer. In Proceedings of the 2017 IEEE Conference on Computer Vision and Pattern Recognition (CVPR), Honolulu, HI, USA, 21–26 June 2017; pp. 3730–3738. [CrossRef]

21. Johnson, J.; Alahi, A.; Fei Fei, L. Perceptual Losses for Real-Time Style Transfer and Super-Resolution. In *Lecture Notes in Computer Science, Proceedings of the Computer Vision—ECCV 2016, Amsterdam, The Netherlands, 11–14 October 2016*; Leibe, B., Matas, J., Sebe, N., Welling, M., Eds.; Springer: Cham, Switzerland, 2016; Volume 9906. [CrossRef]
22. Tan, W.R.; Chan, C.S.; Aguirre, H.E.; Tanaka, K. ArtGAN: Artwork synthesis with conditional categorical GANs. In Proceedings of the 2017 IEEE International Conference on Image Processing (ICIP), Beijing, China, 17–20 September 2017; pp. 3760–3764. [CrossRef]
23. Tan, W.R.; Chan, C.S.; Aguirre, H.E.; Tanaka, K. Improved ArtGAN for Conditional Synthesis of Natural Image and Artwork. *IEEE Trans. Image Process.* **2019**, *28*, 394–409. [CrossRef] [PubMed]
24. Choi, D.; Shallue, C.; Nado, Z.; Lee, J.; Maddison, C.; Dahl, G. On Empirical Comparisons of Optimizers for Deep Learning. *arXiv* **2020**, arXiv:1910.05446.
25. Kingma, D.; Ba, J. Adam: A Method for Stochastic Optimization. In Proceedings of the 3rd International Conference on Learning Representations, ICLR 2015, San Diego, CA, USA, 7–9 May 2015.
26. Daskalakis, C.; Ilyas, A.; Syrgkanis, V.; Zeng, H. Training GANs with optimism. In Proceedings of the International Conference on Learning Representations, Vancuver, BC, Canada, 30 April–3 May 2018. Available online: https://openreview.net/forum?id=SJJySbbAZ (accessed on 20 September 2021).

 *technologies*

Article

Encoding Two-Qubit Logical States and Quantum Operations Using the Energy States of a Physical System

Dimitrios Ntalaperas * and Nikos Konofaos

Department of Informatics, Aristotle University of Thessaloniki, 54124 Thessaloniki, Greece; nkonofao@csd.auth.gr
* Correspondence: ntalaperas@csd.auth.gr

Abstract: In this paper, we introduce a novel coding scheme, which allows single quantum systems to encode multi-qubit registers. This allows for more efficient use of resources and the economy in designing quantum systems. The scheme is based on the notion of encoding logical quantum states using the charge degree of freedom of the discrete energy spectrum that is formed by introducing impurities in a semiconductor material. We propose a mechanism of performing single qubit operations and controlled two-qubit operations, providing a mechanism for achieving these operations using appropriate pulses generated by Rabi oscillations. The above architecture is simulated using the Armonk single qubit quantum computer of IBM to encode two logical quantum states into the energy states of Armonk's qubit and using custom pulses to perform one and two-qubit quantum operations.

Keywords: two-qubit states; quantum computers; quantum gates; Rabi oscillations; single qubit quantum computer; quantum operations simulation

Citation: Ntalaperas, D.; Konofaos, N. Encoding Two-Qubit Logical States and Quantum Operations Using the Energy States of a Physical System. *Technologies* **2022**, *10*, 1. https://doi.org/10.3390/technologies 10010001

Academic Editor: Francesco Plastina

Received: 12 November 2021
Accepted: 17 December 2021
Published: 22 December 2021

Publisher's Note: MDPI stays neutral with regard to jurisdictional claims in published maps and institutional affiliations.

1. Introduction

Quantum algorithms are known to outperform their classical counterparts in a variety of computational tasks [1–3]; various proposals have also been suggested for the physical implementation of quantum computers, while multiple implementations have also taken place [4,5]. Most of the proposed implementation techniques rely on the representation of the quantum logical unit of information, the qubit, to a degree of freedom of the underlying physical system. There are thus qubits that have energy eigenstates as basis states, qubits that have spin eigenstates as basis states and so on. Although a typical quantum computing architecture may refer to fundamental configurations of qubits and their interaction, resulting implementations will normally involve error correcting mechanisms to accommodate for the various sources of quantum error such as measurement error, decoherence and depolarization. While classically, it is straightforward to copy the state of a bit to multiple bits and use redundancy for error correction, the no-cloning theorem prohibits this approach for qubits. Instead, the codes used for qubits involve entanglement; the bit flip code and the Shor Code [6] are typical examples of quantum error correction codes. The need for error correction imposes the necessity of implementing an extra number of qubits to any implementation of a quantum computing architecture. The totality of the qubits used to both store and perform quantum error correction is typically referred to as the physical qubits of the system. In contrast, the logical amount of information encoded in the system is referred to as the logical qubits. Therefore, a system implementing the bit flip code that uses two extra qubits to perform error correction to a single state will have three physical qubits and one logical.

In this paper, we investigate a way to reduce the amount of error, and thus the need to correct extra physical qubits, by investigating the principles of quantum error correction and performing them to the domain of the architecture, i.e., the domain of the logical computation itself. Instead of using the eigenstates of two-state physical systems to encode single qubits, we allow more eigenstates to be used, thus encoding multiple qubits to the eigenstates of a single system. This transforms quantum operations involving multiple qubits, such as the CNOT, to quantum operations that are performed using the eigenstates of a single physical system. In this sense, entangled states at the logical level can be created by involving a single system at the physical level. Since a main source of error is the interactions between qubits, this process is expected to reduce the overall amount of error.

The paper is structured as follows: In Section 2, we give a brief overview on how, in principle, multiple quantum logical states can be mapped to single physical states with an example of how a CNOT gate can be implemented using a harmonic oscillator. We then build upon the main idea of the harmonic oscillator to propose a more robust schema of encoding information using the charge degree of freedom of impurity atoms embedded in semiconductor materials. Section 3 provides the core idea by presenting how the mapping model is implemented in charge qubits defined by donor electrons of impurity atoms embedded in a semiconductor structure. The method by which single and two-qubit quantum gates are implemented in the model is also presented in Section 3. where, we also perform a simulation of the proposed architecture using the IBM Armonk single qubit computer [5]. Finally, Section 4 offers a discussion of the results of this work.

2. Materials and Methods

The idea of encoding multiple qubit states to a single physical state is not new. An example can be seen in the quantum oscillator case [7]. Though not a good candidate for the physical realization of quantum physical systems due to issues with scaling and equidistant energy separation, the quantum oscillator can be used to demonstrate the principle of mapping multiple qubit states to a single physical state.

Consider, for example, the following mapping:

$$
\begin{aligned}
|00\rangle_L &\rightarrow |0\rangle \\
|01\rangle_L &\rightarrow |2\rangle \\
|10\rangle_L &\rightarrow \frac{(|4\rangle + |1\rangle)}{\sqrt{2}} \\
|11\rangle_L &\rightarrow \frac{(|4\rangle - |1\rangle)}{\sqrt{2}}
\end{aligned}
\tag{1}
$$

where subscript L corresponds to the logical quantum states. States on the right side, appearing without subscript, correspond to the energy levels of the quantum oscillator. The energy eigenstates of the quantum oscillator evolve with time as:

$$
|n\rangle \rightarrow e^{-i\pi n\omega t}|n\rangle
\tag{2}
$$

Consequently, if we allow the system to evolve for time t equal to $\pi/\hbar\omega$ then the state changes according to:

$$
|n\rangle \rightarrow (-1)^n|n\rangle
\tag{3}
$$

so that odd labeled physical states (in this case, the $|1\rangle$ state) change sign. Applying for the logical states appearing in the Equation (1) state $|10\rangle$, changes to $|11\rangle$ and vice versa, while the other two states remain unchanged, corresponds to the truth table of the quantum CNOT gate; considering the first qubit to be the control and the second one the target qubit.

Although the quantum harmonic oscillator architecture presented above is more of a theoretical schema, architectures that divert from the typical mapping of a single physical degree of freedom to a single qubit state have been developed. A hybrid approach, for example, that uses different mechanisms to encode different parts of a quantum computation has been implemented in [8], where the control qubit of a C-SWAP Gate was implemented using a photon's polarization, whereas the SWAP part was implemented using four degrees of the photonic angular momentum. In general, the concept of hyperentanglement [9] allows the usage of multiple degrees of freedom of a single physical system to encode quantum information. The approach presented in this paper uses only one degree of freedom.

The typical spin systems used for quantum computing consist of two eigenstates sufficient for encoding a single qubit. For the purpose of the current work, we needed to map multi-qubit registers to eigenstates of single physical systems; therefore, charge qubits are a better candidate that allow for such encodings. Charge qubits are implemented using the energy charge of freedom of a quantum system. Typical efforts to construct an architecture based on charge qubits include the encoding of a qubit based on the presence of Cooper pairs [10] or encoding to the charge degree of freedom of electrons in semiconductor devices [11,12]. More recently, architectures that are based on neutral atoms have been proposed [13]. These architectures make use of laser beams targeted at an atom ensemble to cool them to temperatures of the order of mK; they then use pulses to excite the atoms and use the energy states as the computational basis.

Similar to the two-spin system is the archetypal model for spin qubits, in which a two-level atom can be used for prototyping quantum logic operations on a single qubit. Scaling a physical system to allow for operations on an arbitrary number of qubits is one of the biggest challenges of quantum computing. In the case of the two-level atom, the straightforward method of scaling the system by adding more energy levels becomes quickly inefficient, as the gap between higher levels becomes very narrow and, after a certain point, practically continuous. This is also the case for the harmonic oscillator state presented above.

This difficulty can be overcome by various methods such as those mentioned in the first paragraph; for the purposes of the current work, a semiconductor material with a pentavalent donor impurity contributing one extra donor electron is considered (Figure 1). Depending on the material and the doping substance, multiple energy eigenstates may be introduced in the semiconductor bandgap. For the purposes of the present treatment, it will be assumed that energy levels can be raised as desired by the appropriate placement of impurities in the semiconductor grid and that any degeneracies can be lifted by applying the appropriate external electric fields. It is further assumed that for the transitions of interest, the selection rules resulting from the symmetries of the material are always allowed and that energy levels may always be defined such that the energy differences between transition levels may always be matched by incoming electromagnetic pulses. All other energy levels are sufficiently detuned and thus not affected by the incoming pulse.

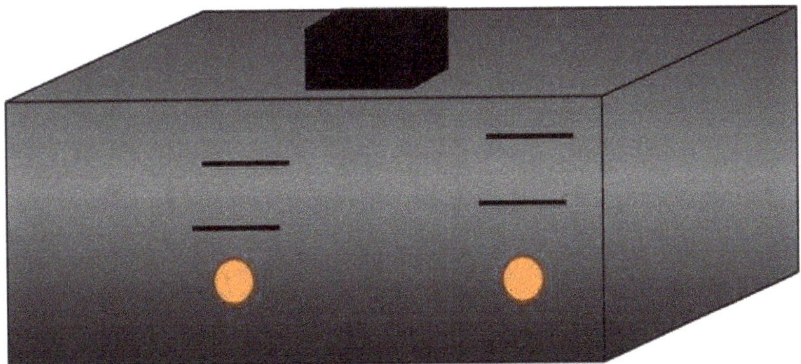

Figure 1. Basic setup of an impurity pair of atoms embedded in a semiconductor.

For the pair energy levels depicted in Figure 1, the following mapping between physical and quantum logical states is considered:

$$|00\rangle \rightarrow |1\rangle$$
$$|01\rangle \rightarrow |2\rangle$$
$$|10\rangle \rightarrow |3\rangle \tag{4}$$
$$|11\rangle \rightarrow |4\rangle$$

Transitions between states will be analyzed for two cases, namely single qubit and two-qubit operations.

A typical method for a quantum state, to make transitions between energy eigenstates, is by making use of Rabi oscillations. The dynamics of Rabi oscillations have already been studied as candidates for encoding qubits and manipulating quantum information [14,15]; for the domain of semiconductors in particular, there are already results that demonstrate the feasibility of quantum control using spin qubits in GaAs quantum dots [16] and Si quantum dot systems [17]. Here we will give an overview on how the same approach can also be followed to manipulate the logical states, as these were defined in Equation (4).

Consider that an electric pulse is applied to the electron in Figure 1. It can be shown that under the rotating wave approximation [18], the state of the electron will evolve according to:

$$\left[\dot{c}_0(t)\ \dot{c}_1(t)\right] = \left[\frac{i\Omega_R}{2}e^{-i\varphi}e^{\Delta t}c_0(t)\ \frac{i\Omega_R}{2}e^{-i\varphi}e^{\Delta t}c_1(t)\right] \tag{5}$$

c_i is the amplitude of the *ith* state, Ω_R the Rabi frequency determined by the energy of the incident pulse and the electric dipole matrix and $\Delta = \nu - \omega$ the detuning, ν is the frequency of the incident pulse and $\omega = \omega_2 - \omega_1$ is the frequency difference between the two states. An electric field that is near-resonant with the transition when applied for a finite time, can be used to perform an arbitrary rotation between the two states, the exact time being dependent on the detuning and Rabi frequency.

Consider the setup depicted in Figure 2, where $\omega_{1\to2} = \omega_{3\to4}$ with $\omega_{1\to3}$ and $\omega_{2\to4}$ being sufficiently detuned that a pulse nearly resonant with $\omega_{1\to2} = \omega_{3\to4}$ is applied. This pulse must be applied for sufficient time to perform any transition between states $|1\rangle$ and $|2\rangle$ if the electron is in one of these states, or between $|3\rangle$ and $|4\rangle$ if the electron is in one of those states. The mapping defined in Equation (4) corresponds to a single qubit operation on the second qubit. If a single qubit operation is required in the first qubit, the same argument may be followed for frequencies $\omega_{1\to2} = \omega_{3\to4}$, which will allow the first qubit to be arbitrarily controlled.

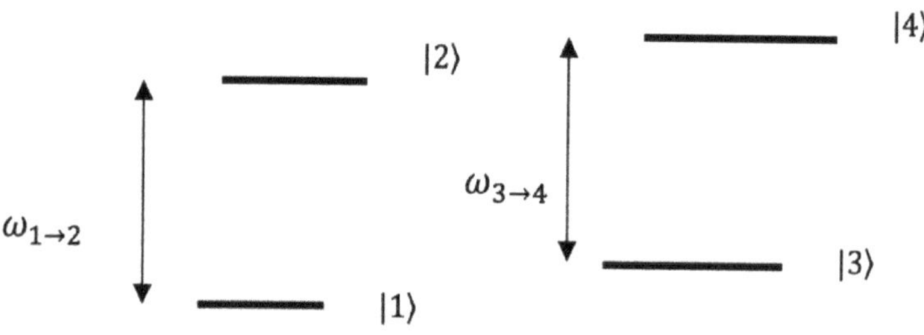

Figure 2. Basic setup for the one-qubit operation.

Though the above setup will only work for frequencies that are pairwise degenerate, the technique may be slightly modified to accompany non-degenerate states, providing those transitions are allowed by the selection rules and that all frequencies are sufficiently detuned. The modification is implemented to perform two pulses in succession; the detuned pulse will not affect the electron, while the resonant pulse will perform the desired rotation.

For the two-qubit case, based on Equation (4), and considering the rightmost logical qubit to be the controlled qubit, the required transitions for the CNOT gate at the physical representation are:

$$\begin{aligned}
|1\rangle &\to |1\rangle \\
|2\rangle &\to |2\rangle \\
|3\rangle &\to |4\rangle \\
|4\rangle &\to |3\rangle
\end{aligned} \tag{6}$$

If $\omega_{3\to4}$ is sufficiently detuned from any other transition frequency, a π Rabi pulse with this resonance will perform the desired transition.

3. Results

To establish the validity of the approach, we exhibited the dynamics of the system using the IBM Armonk, a single qubit quantum computer developed by IBM that can be accessed via the Qiskit SDK [5]. For the purpose of this work, encoded four quantum states into the physical states of Armonk's single qubit. Though Armonk provides its own set of single qubit gates, we drove it using custom-made pulses, making use of its higher energy states. The possibility of using higher energy levels of a transmon system has already been studied in [19], where the authors proved, by studying decoherence and decay times for π pulses used for consecutive excitation frequencies, that higher excited states could be used for computation.

In this work, for our proposed mapping, treated two separate cases: (a) Single qubit operations and (b) two-qubit operations

3.1. Single Qubit Operations

Single qubit operations involve the generation of pulses that, when applied to the system, alter only one of the qubits in the logical space. In a similar fashion, with the encoding of multiple qubit states in the charge degree of freedom of the electron of an impurity atom embedded in semiconductor material explained in Section 2, let us consider energy states $|0\rangle$, $|1\rangle$, $|2\rangle$, $|3\rangle$ of Armonk's single qubit and follow a similar mapping, as in Equation (4):

$$
\begin{aligned}
|00\rangle_L &\to |0\rangle_{Armonk} \\
|01\rangle_L &\to |1\rangle_{Armonk} \\
|10\rangle_L &\to |2\rangle_{Armonk} \\
|11\rangle_L &\to |3\rangle_{Armonk}
\end{aligned}
\tag{7}
$$

We investigated how we can create pulses that, when performed at the level of physical qubits, isolate qubits at the logical level. Taking the logical ground state $|00\rangle_L$ for example, creating a superposition of equal probabilities for the first qubit will take the state to $\frac{1}{\sqrt{2}}(|00\rangle_L + |01\rangle_L)$, which, as can be seen by Equation (7), is equivalent to the physical state $\frac{1}{\sqrt{2}}(|0\rangle_{Armonk} + |1\rangle_{Armonk})$. To drive this, we needed to create a pi/2 pulse between the ground and the first excited state of the physical qubit. This was achieved using IBM's Quantum Experience Aer library [5]. Briefly, the steps followed were:

1. Calibration of the qubit to derive the transition frequency.
2. Derivation of the characteristics of the pi Rabi pulse.
3. Confirmation of the Rabi pulse by applying it to the ground state of Armonk's qubit.
4. Clustering the samples of the measured results and an equivalent number of ground states to derive the mean values around the ground and first excited states.
5. Based on the characteristics of the Rabi pulse derived in Step 3, define a pi/2 pulse with half the amplitude of the original Rabi pulse used to perform a full flip.
6. Drive the Armonk's qubit originally set to the ground state, with the pulse derived at Section 3.1 and measure the results.

Figures 3 and 4 show a scatter plot with the results obtained by performing the above steps in Armonk's qubit 1024 times. The two black dots correspond to the mean values of the ground and first excited states, as these were derived at Step 4 after calibrating the qubit. It can be seen that our results are roughly equally separated between the ground and first excited state, as was expected from the mappings discussed above.

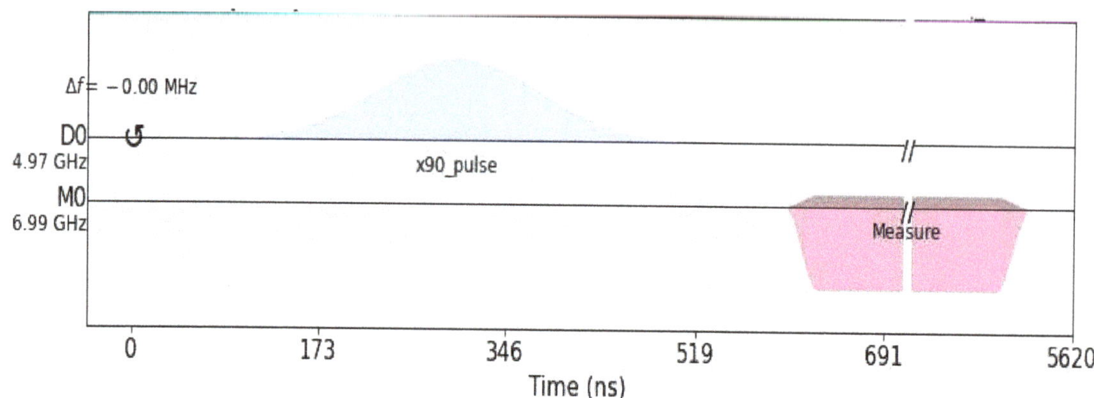

Figure 3. Calibrated pulse for pi/2 rotation of Armonk's qubit.

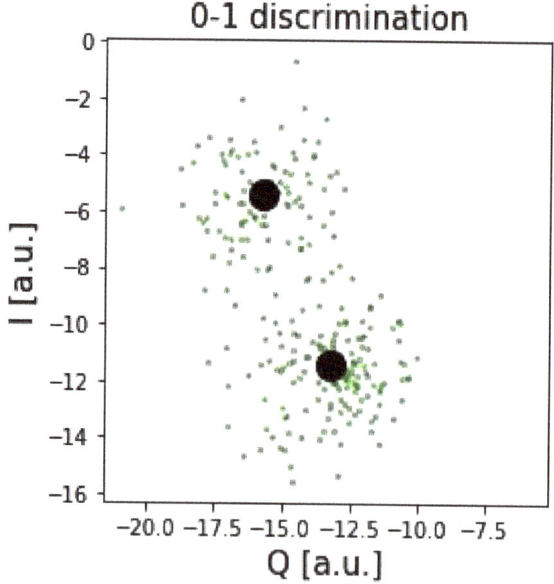

Figure 4. Scatter plots of the results obtained by driving Armonk's qubit with the pi/2 pulse.

Although the above discussion depicts the feasibility of isolating qubits at the logical level, for a general single qubit gate, the full truth table should be taken into account when constructing pulses. The truth table of the Hadamard Gate, for example, when considering both states of the logical qubit that does not take part in the computation, can be summarized as:

$$
\begin{aligned}
|00\rangle_L\, H_1 &\rightarrow |0\rangle \otimes \tfrac{1}{\sqrt{2}}(|0\rangle+|1\rangle)_L = \tfrac{1}{\sqrt{2}}(|00\rangle+|01\rangle)_L \rightarrow \tfrac{1}{\sqrt{2}}(|0\rangle+|1\rangle)_{Armonk} \\
|01\rangle_L\, H_1 &\rightarrow |0\rangle \otimes \tfrac{1}{\sqrt{2}}(|0\rangle-|1\rangle)_L = \tfrac{1}{\sqrt{2}}(|00\rangle-|01\rangle)_L \rightarrow \tfrac{1}{\sqrt{2}}(|0\rangle-|1\rangle)_{Armonk} \\
|10\rangle_L\, H_1 &\rightarrow |1\rangle \otimes \tfrac{1}{\sqrt{2}}(|0\rangle+|1\rangle)_L = \tfrac{1}{\sqrt{2}}(|10\rangle+|11\rangle)_L \rightarrow \tfrac{1}{\sqrt{2}}(|2\rangle+|3\rangle)_{Armonk} \\
|11\rangle_L\, H_1 &\rightarrow |1\rangle \otimes \tfrac{1}{\sqrt{2}}(|0\rangle-|1\rangle)_L = \tfrac{1}{\sqrt{2}}(|10\rangle-|11\rangle)_L \rightarrow \tfrac{1}{\sqrt{2}}(|2\rangle-|3\rangle)_{Armonk}
\end{aligned}
\tag{8}
$$

Therefore, it is necessary to also consider the higher energy states when constructing a full Hadamard Gate. Following the same approach as before, we constructed a pi/2 pulse between Armonk's higher energy states of $|2\rangle$ and $|3\rangle$. Section 3.2 discusses how to construct such pulses, as these are needed for the two-qubit operations with the mappings used. Assuming that the pulses used to drive transitions between $|0\rangle$ and $|1\rangle$, and between $|2\rangle$ and $|3\rangle$ are sufficiently detuned, as is the case for a correctly calibrated qubit, a combined pulse that consists of pi/2 pulses for each one of the transitions will be able to create superpositions of the first logical qubits, regardless of the state of the other qubit.

3.2. Two-Qubit Operations

Recalling the discussion in Section 3.1 and applying it to the Armonk qubit, the correspondence between logical and physical states for the CNOT Gate can be summarized as:

$$\begin{aligned}
|0\rangle_{Armonk} &\rightarrow |00\rangle_L \; CNOT \rightarrow |00\rangle_L \rightarrow |0\rangle_{Armonk} \\
|1\rangle_{Armonk} &\rightarrow |01\rangle_L \; CNOT \rightarrow |01\rangle_L \rightarrow |1\rangle_{Armonk} \\
|2\rangle_{Armonk} &\rightarrow |01\rangle_L \; CNOT \rightarrow |11\rangle_L \rightarrow |3\rangle_{Armonk} \\
|3\rangle_{Armonk} &\rightarrow |11\rangle_L \; CNOT \rightarrow |10\rangle_L \rightarrow |2\rangle_{Armonk}
\end{aligned} \tag{9}$$

It is thus sufficient to construct a pulse that performs a flip between states $|2\rangle$ and $|3\rangle$ of the Armonk's qubit. Unfortunately, there is no direct way of exciting higher energy states from the ground state of Armonk's qubit due to the limitations of maximum pulse power. To overcome this, we used the first excited state (which can be reached by applying a pi pulse to the ground state in a similar fashion as explained in Section 3.1) and then applied a sideband to this base pulse. We then gauged the qubit for a response and retrieved the candidate frequency for the 1→2 transition. After this, we can fully calibrated the other characteristics of the transition pulse (e.g., amplitude) following the same procedure as in Section 3.1. The definition of the 2→3 pulse followed the same rules; we excited the state from 0 to 1 and then to 2, after this, we applied another sideband to find the transition frequency of 2→3. Figure 5 depicts the definition of the sweeping pulse (main plus sideband) for the 1→2 transition. In contrast, Figure 6 depicts the measured signal for each of the test frequencies. For this particular experiment, the first two extrema, centered at around 2.63 GHz, were considered for constructing and testing the Rabi pulse for the transition.

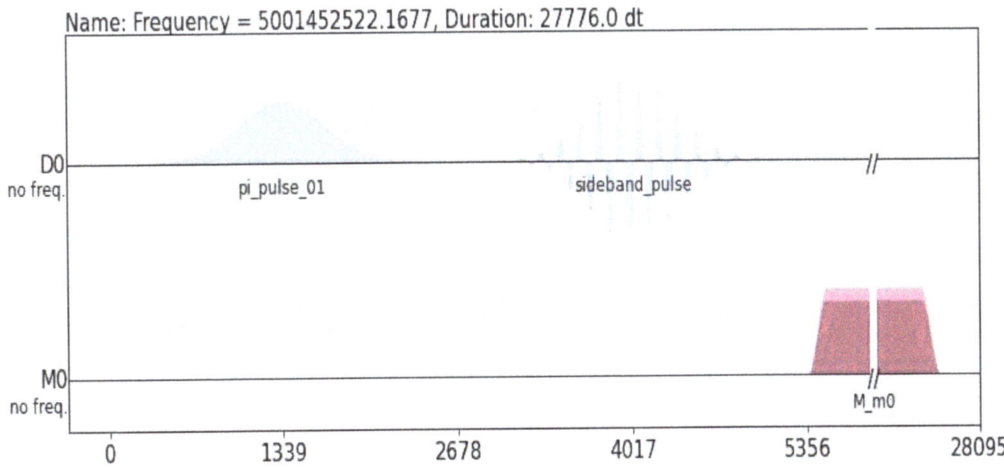

Figure 5. Main and sideband pulse for sweeping for the 1→2 transition frequency.

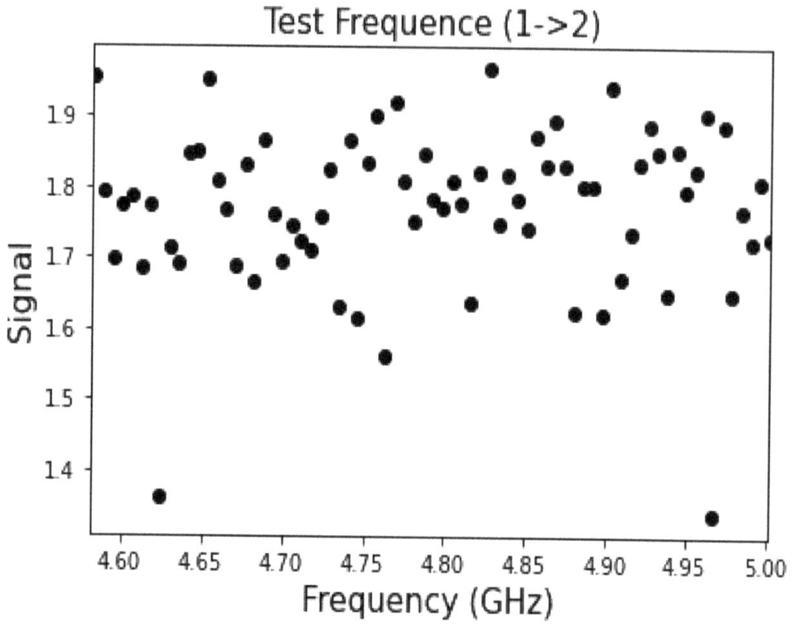

Figure 6. Signal measurement for the measured transitions. The first candidate extremum appears at ~4.6 GHz.

It is to be noted that, as the level of the excited state grows, experiments may become less reliable and dependent upon the calibration status of the system. However, they can be used to depict that, in principle, higher-level energy states can be used to perform encodings involving multiple logical quantum states.

Concerning the error propagation, we can estimate that each transition required induces an error that is of the order of the error for the Pauli X Gate of the architecture. For the Armonk quantum computer, this is equal to 3.27×10^{-4} A single logical qubit operation requires two such pulses, whereas the controlled operation requires only one. Assuming accumulative error and no extra error correction between the steps of computation, Table 1. depicts the total number of gates supported to achieve a computation with an accumulated error of less than 10%. It should be noted as the scaling factor increases (more logical qubit states are encoded into distinct energy levels), operations will involve more pulses, thereby further decreasing the circuit depth.

Table 1. Estimation of total number of Hadamard and CNOT Gates that can be performed on Armonk for the two-qubit case.

Total Circuit Depth	# of Hadamard Gates	# of CNOT Gates
300	0	300
157	157	0
228	76	152

4. Discussion

In this paper, we investigated the possibility of using multiple energy states to densely encode logical qubit states. We pointed out a possible physical system where this architecture can be implemented; this consists of a semiconductor device embedded with impurities. These systems are generally robust enough to introduce multiple energy bands, as these can be controlled by the number and types of impurities. We finally demonstrated the physical possibility of manipulating such states by encoding a two-qubit register into the first four energy states of IBM's single qubit computer Armonk, and constructing specific pulses consistent with the logical truth tables of the quantum Hadamard and Controlled NOT gates. One interesting feature that was observed was the fact that for quantum gates that involve more than one qubit, the required physical manipulation was less complex than that required for the single qubit case. This behavior is to be expected, as altering the physical state of the underlying physical qubit may induce changes to multiple logical basis states. In principle, the more constraining the logical operation is (e.g., CNOT), the less manipulation of energy levels it will require.

Scaling the above schema to more dense coding is straightforward; however, the physical limitation may put an upper bound in the number of states encoded by a single physical system, with the exact bound being dependent on the underlying architecture. One limitation is the energy separation. Although we can, in theory, envisage introducing multiple controllable energy states for donor atoms, there is a limit where separating these states may become practically impossible. Another limitation is the number of pulses needed to perform quantum logical gates, which scales up as the number of logical quantum states that are encoded into the energy spectrum increases. For the case equal to two that was studied in the present work, we needed two pulses for single-qubit gates and one pulse for two-qubit gates; it is straightforward to see that for the mapping of a three logical state, we would need three pulses of a single-qubit gate, two pulses of a two-qubit case and we could perform a three-qubit case (e.g., Toffoli gate) with a single pulse. In general, the amount of single-system pulses grows linearly with the amount of qubits encoded. The extent to which our approach is efficient depends on the parameters of the underlying architecture. For example, for the transmon case, the Belem backend of IBM has an average error of an X gate equal to $\sim 2.5 \times 10^{-4}$, while that of the CNOT is $\sim 1 \times 10^{-4}$. The relaxation and dephasing times are equal to ~75 us and ~90 us respectively. Though these parameters are computed for the base case, where only the ground and first excited state are used and, therefore, need experimental re-evaluation for the case of higher excited states used in our work, it can be seen that in first-order, they hint that the number of single-qubit operations that can be performed before the error accumulated is comparable to that of the controlled case, is in the order of ~100.

It should be noted that even with systems specifically designed to accommodate higher energy levels into the computation, it cannot realistically be expected that a complete circuit that is of a size sufficient to perform useful computation can be performed in its entirety in the energy spectrum of a physical system. By separating qubits into logical groups, within which only single system pulses are needed, we expect that the overall number of controlled pulses will be significantly reduced. Nevertheless, physical coupling operations will still be required to perform logical controlled operations between qubits that belong to different groups. Future work will focus on handling these issues by introducing couplings between neighboring physical systems; proof of the concept is being developed using Belem, a 5-qubit backend provided by IBM, with the aim of using the results to expand the semiconductor model.

Author Contributions: Conceptualization, and methodology, D.N.; software, D.N.; validation, D.N. and N.K.; formal analysis, D.N.; investigation, D.N.; resources, D.N. and N.K.; data curation, D.N.; writing—original draft preparation, D.N.; writing—review and editing, N.K.; visualization, D.N. and N.K.; supervision, N.K., project administration, N.K.; funding acquisition, D.N. and N.K. All authors have read and agreed to the published version of the manuscript.

Funding: This research received no external funding.

Data Availability Statement: All data are available to any researcher upon request.

Conflicts of Interest: The authors declare no conflict of interest.

The problem becomes harder to solve when considering that the introduced timing violations may involve multiple corners that may need significantly different actions to remove them.

The least disruptive operations for improving design's characteristics during physical synthesis involve threshold voltage (V_T) reassignment and gate sizing [4,5]. V_T reassignment tradeoffs smaller delay with increased leakage power and does not perturb routing nor it requires a new parasitics extraction after the change. Gate resizing, even if not as simple as V_T reassignment, is still considered a fairly noninvasive operation. In the worst case, increasing cell's size (possibly avoiding exceedingly large changes) may require an additional local legalization step [6,7] and local rerouting of certain nets [8].

Inserting buffers is still an option at this step [9–11]. However, buffer insertion may ruin local placement and routing, which may be hard to fix later in highly congested designs. Other highly powerful optimization steps such as useful clock skewing are also considered hard to apply at the end of the flow, unless there is no other practical way to solve the remaining timing violations [10,12,13].

Gate sizing and V_T assignment algorithms have a long history in physical synthesis flows. Initial works assumed continuous sizes for the gates [14] but these approaches had delay inaccuracies compared to that of the real discrete gate sizes [15]. Coudert et al. [16] was from the first ones that proposed a gate sizing method that handles such discrete sizes. Many different methods were studied to solve the size selection problem effectively. For example, linear programming (LP) was used widely in the literature [17–20]. Simulated annealing was also used to solve the gate sizing problem because it can be applied on circuits containing million gates [21]. Daboul et al. [22] used the formulation of resource sharing to select gate sizes. Other approaches proposed to apply dynamic programming (DP) [23–26]. Alternative works use sensitivity functions, and from the available sizes, select the size that maximizes the power reduction with the minimal timing degradation [27,28]. Some of these works were extended to handle multiple timing corners and scenarios for more realistic designs [3,29]; even machine learning was used for gate sizing. The latest work of [30] uses deep reinforcement learning to change the sizes and shows high-quality final results.

Among the large set of available solutions, those that rely on Lagrangian Relaxation (LR) achieve significantly better result [15,31–35]. However, when applied incrementally they need many iterations to converge even if the number of timing violators is small. Most LR-based sizers assume that they are allowed to initialize every cell of the design to a chosen initial state, e.g., initialize all cells to their minimum size [36], before beginning the optimization. This design disruption may seem reasonable at the early steps of the flow but is not allowed close to the the the end.

In this work, we propose a novel initialization strategy for multicorner LR-based timing/power optimizers across multiple operating conditions that combines two useful benefits: on one hand, we enjoy the optimization efficiency of an LR-based gate sizer and on the other hand we enjoy fast runtimes and true incremental operation, i.e., the optimized design is only marginally different from the original design, but with the timing violations of multiple corners repaired.

The proposed approach was compared to a fully-fledged LR-based gate sizer on optimized versions of the benchmarks of the ISPD2013 contest [37] across single and multiple corners by enhancing the work in [38]. The used benchmarks experience small timing violations due to local changes of their routed wires. In both single and multicorner cases, the proposed initialization strategy successfully optimizes the timing with reduced runtime. Each design has 37% better timing performance on average with reduced leakage power and the runtime is reduced by more than 43% on average, since it simplifies the convergence of the algorithm.

2. Basics of LR-Based Gate Sizing

A timing-driven optimizer tries to minimize the power (or area) of the design given a set of timing constraints.

$$\text{minimize} \quad \sum_i leakage_i \tag{1}$$

$$\text{subject to} \quad a_i + d_{ij} \leq a_j \quad \forall\, i \to j$$

$$a_k \leq r_k \quad \forall \text{ endpoints } k$$

Variable a_i denotes the arrival time at the output pin of cell i while $d_{i,j}$ is the sum of wire and cell delay of the timing arc $i \to j$ which is defined from the output pin of the gate i to the output pin of the gate j. Figure 1 depicts the delays involved in the computation of d_{ij} for different cases. For combinational gates in the middle of a logic netlist, as shown in Figure 1a, d_{ij} is the summation of the wire delay and the gate delay from output of gate i to the output pin of gate j.

Pin j may represent also a timing endpoint. Timing endpoints can be the primary outputs (POs) of a design or the inputs of flip-flops. When pin j belongs to the set of primary outputs (POs), as highlighted in Figure 1b, delay d_{ij} is equal to the wire delay connecting the output pin of driver i and primary output j. Similarly, when pin j is a flip-flop input, shown in Figure 1c, delay d_{ij} involves only the wire delay from driver i to the input D-pin of the flip-flop j. Parameter r_k is the required arrival time at any timing endpoint k [39].

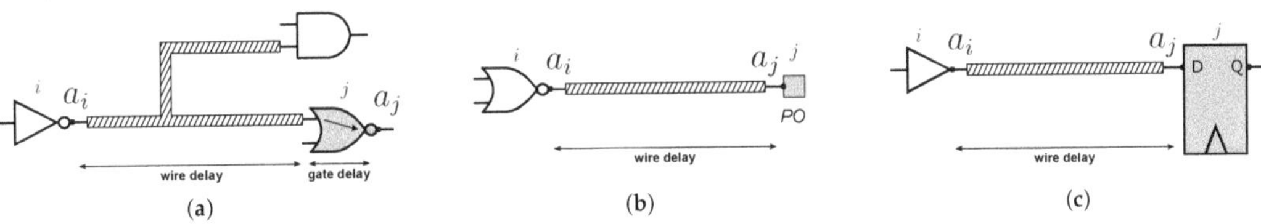

Figure 1. Definition of arrival times a_i, a_j and delay $d_{i,j}$ for (**a**) combinational gates, (**b**) primary outputs and (**c**) flip-flops.

Associating the constraint for each timing arc with a non-negative Lagrange Multiplier (LM) λ_{ij}, that acts as a penalty factor when the respective constraint gets violated, and computing the KKT optimality conditions [15,40,41], allows us to simplify the constrained minimization problem (1) to the equivalent unconstrained minimization problem (2).

$$\text{minimize} \quad \sum_i leakage_i + \sum_{i \to j} \lambda_{ij} d_{ij} \tag{2}$$

The KKT optimality conditions with respect to the values of LMs impose that Equation (3) should hold during optimization for all pins of the design

$$\sum_{i \to j} \lambda_{ij} = \sum_{j \to k} \lambda_{jk} \tag{3}$$

For the example shown in Figure 2, Equation (3) for gate 6 implies that $\lambda_{36} + \lambda_{46} = \lambda_{67} + \lambda_{68}$.

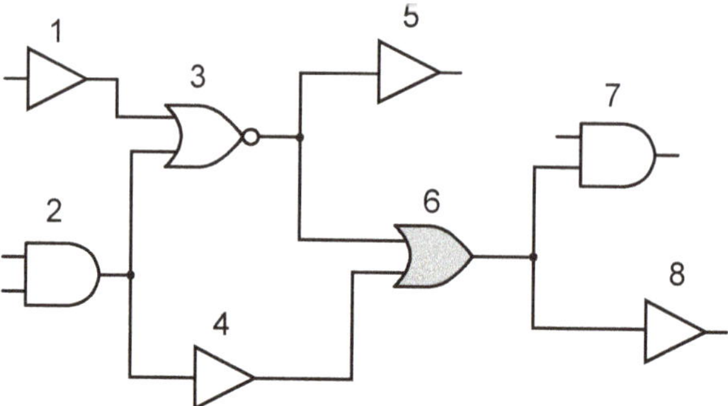

Figure 2. Example design to show which LMs are considered in computation of local cost and to present LM equalities to preserve optimality KKT conditions, which imply that sum of input LMs must be equal to sum of output LMs. Thus, for highlighted gate 6, LMs $\lambda_{13}, \lambda_{23}, \lambda_{24}, \lambda_{35}, \lambda_{36}, \lambda_{46}, \lambda_{67}$ and λ_{68} multiplied with their corresponding delays will form local cost. For LM propagation, for gate 6, we should guarantee that $\lambda_{36} + \lambda_{46} = \lambda_{67} + \lambda_{68}$.

State-of-the-art LR-based optimizers [15,31–34] try to minimize the global cost function (2) using many iterations of local gate resizing and V_T reassignment steps. The overall optimization flow is depicted in Figure 3.

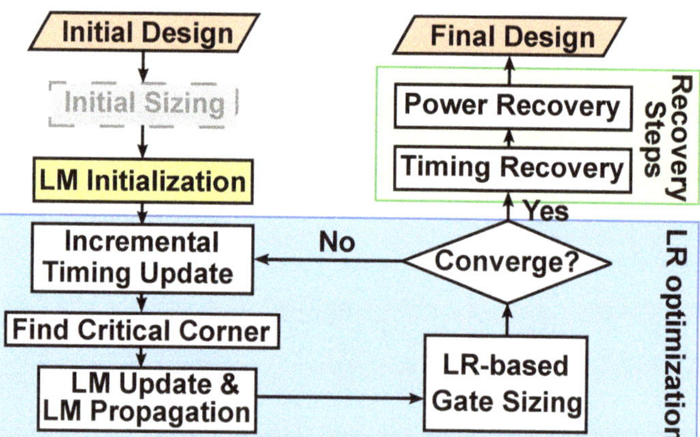

Figure 3. Overall LR-based gate sizing optimization flow.

Initially, all gates are downsized to their least leakage power option (lowest size and highest V_T) [36,42] that does not violate any design rule constraint. Solving design-rule violations early, simplifies the following local logic tuning steps. In the following, all LMs are set to a starting value, usually to 1, and the main LR optimization loop starts. Each iteration of LR-based gate sizing begins with a full incremental timing update and then evolves in two phases. In the first phase, the LMs are updated and propagated to all gates to reflect the new criticality of the corresponding timing arcs. In the second phase, for each gate, examined in topological order, all possible discrete cell sizes and threshold voltages are tried, assuming constant LMs. The new version selected for the resized gate is the one that minimizes the cost function (2) and does not introduce any design-rule violations.

At each iteration, a full incremental timing update on all examined corners is needed to reflect the new timing violations. From all the available corners, the most critical corner is identified [10] for the current iteration. When there aren't any timing violations in any corner, we name critical the timing corner that gives the lowest total slack.

With the new timing information updated, the LMs should be updated too. The update may take different forms and can be either additive ($\lambda_{\text{new}} = \gamma + \delta\lambda_{\text{old}}$) or multiplicative ($\lambda_{\text{new}} = \gamma\lambda_{\text{old}}$) [43]. Following the proposal of [31] we use a multiplicative LM update depicted in (4):

$$
\begin{aligned}
\lambda_{ij} &= \lambda_{ij}\left(1 + \frac{a_j - r_j}{T}\right)^{1/M} \quad \forall \text{ timing arc } i \to j \text{ with } a_j \geq r_j \\
\lambda_{ij} &= \lambda_{ij}\left(1 + \frac{r_j - a_j}{T}\right)^{-M} \quad \forall \text{ timing arc } i \to j \text{ with } a_j < r_j
\end{aligned}
\tag{4}
$$

Once the LMs were updated, LMs must be propagated from output to input following a reverse topological order. In this way, the timing criticality measured at the timing endpoints should be transferred gradually to the internals gates of the design. LM propagation updates the LM values of internal timing arcs while still respecting KKT conditions (3).

The value of each LM reflects the timing criticality of each timing arc. LMs increase fast for critical timing arcs and reduce for noncritical timing arcs to favor power reduction. Implicitly, LMs keep also historic information (for the lifetime of an optimization run) with respect to the criticality of each timing arc. If a timing arc remained critical for multiple iterations it is still assumed critical by keeping a high value of LM, even if the slack at its output becomes positive in a certain iteration. In this way, drastic oscillations between critical and noncritical timing arcs are avoided and the optimization evolves smoothly reducing power while satisfying timing constraints.

Later on, and assuming constant LMs, all gates are visited in topological order and for each gate the best size is selected using the same procedure described in Algorithm 1. Firstly, the initial size of the gate is stored and then, each equivalent size of the gate is tried. If the new tried size violates any design rule constraint, this size is rejected. Otherwise, the timing is updated locally, recomputing the new delays and slews of all nets that the examined gate is connected to. To avoid timing degradation, sizes that violate timing constraints are also rejected. If not, the local cost is calculated as the summation of the leakage power of the new size and the neighbor arc delays multiplied by their corresponding LMs.

In the local cost, only the arcs whose delay may have changed are included. These are the arcs of the immediate fanin and fanout cells of the examined gate and the arcs of cells driven by the gates fanin cells. Referring to Figure 2, changing the size of the highlighted gate 6, the local cost consists of the arcs of its immediate fanin cells ($1 \to 3, 2 \to 3, 2 \to 4$), its immediate fanout cells ($6 \to 7, 6 \to 8$) and the arcs of gates driven by the fanin cells of gate 6 ($3 \to 5, 3 \to 6, 4 \to 6$). After trying all the equivalent sizes, the size that minimizes the local cost is selected.

The iterative optimization flow stops when the maximum number of iterations is reached or when the Total Negative Slack (TNS) and total leakage power are assumed unchanged. Some timing violations may still remain, if the gate sizing exchanged some marginally positive slack to further reduce the power. The timing recovery step that follows will solve these violations resizing only specific gates that affect many timing endpoints. For these gates, only the next bigger size is tried and full incremental timing update is performed. Once the timing is closed, the final power recovery step will try to save leakage power without creating new timing violations. Again, each gate is resized only to its either next smaller size or exact higher V_T and an incremental timing update is performed after each try, to have the accurate timing information.

Algorithm 1: Find best size for gate *g*.

```
1  min_cost ← inf ;
2  best_size ← size(g) ;
3  foreach equivalent size s of g do
4  │    resize g to s ;
5  │    if violates_Design_Rule_Constraint(g) then
6  │    │    skip s ;
7  │    end
8  │    update_timing_locally(g) ;
9  │    if timing_degradation_around(g) then
10 │    │    skip s ;
11 │    end
       // Using Equation (2)
12 │    cost ← leakage_g + Σ_{i→j around g} λ_ij d_ij ;
13 │    if (cost < min_cost) then
14 │    │    min_cost ← cost;
15 │    │    best_size ← s;
16 │    end
17 end
18 resize g to best_size ;
19 update_timing_locally(g) ;
```

---

3. Incremental LR-Based Gate Sizing

The overall effectiveness of an LM-based gate sizer is the combined result of the initialization of gate sizes, the strength of the local optimization, and the appropriate update of LMs.

Initializing all cells to their minimum size simplifies the removal of any design rule violations and also may alleviate the design from timing violations because some gates are faster due to the less output load. After initialization, the total leakage power in cost function (2) assumes its minimum value. Thus, the sum of $\lambda_{ij} d_{ij}$ products determine which cell should be selected for each gate. This conclusion holds even if leakage and delay participate normalized to the cost function. Increasing fast the LMs of critical timing arcs guides the optimization to reduce their corresponding delay to minimize their $\lambda_{ij} d_{ij}$ product. As long as timing constraints are not satisfied, LMs keep increasing thus leading to cells with improved delay.

3.1. What Is the Problem?

In an incremental optimization scenario, which is the focus of this work, the first step of the state-of-the-art LR-based gate sizing flow as depicted in Figure 3 cannot be applied. Since the design is almost finalized, the gate sizer is not allowed to "reset" the state of the design and initialize every gate to its minimum size. Therefore, since all gates keep their already decided size the sum of leakage power in (2) may possibly dominate the cost function. The LMs that fit to this occasion are unknown and initializing all of them to an a priori value, e.g., 1, may not be the best choice.

Inevitably, at the first iterations of LR-based gate sizing, lower power cells would be preferred for each gate since they would minimize local cost at the expense of timing. Once timing would starting getting much worse and the corresponding LMs start to take higher values, only then the $\lambda_{ij} d_{ij}$ products would favor the selection of the delay-optimal cells. Due to improper initialization, state-of-the-art LR-based gate sizers exhibit a counterproductive behavior. The less timing critical is the initial state of the design, the more time an LR-based gate sizer would need to optimize it, when resetting the state of the design is not allowed.

To highlight this behavior, we performed an experiment using the pci_bridge32_fast design of the ISPD13 benchmark set. Figure 4 depicts the evolution of the design's Total

Negative Slack (TNS) during LR-based gate sizing for three different cases. When the design suffers from many timing violations (case A), the LR-based gate sizer is able to find fast the way to improve timing, leading to almost closed timing after the first six iterations. The remaining iterations are used to improve leakage power without degrading timing in the meantime.

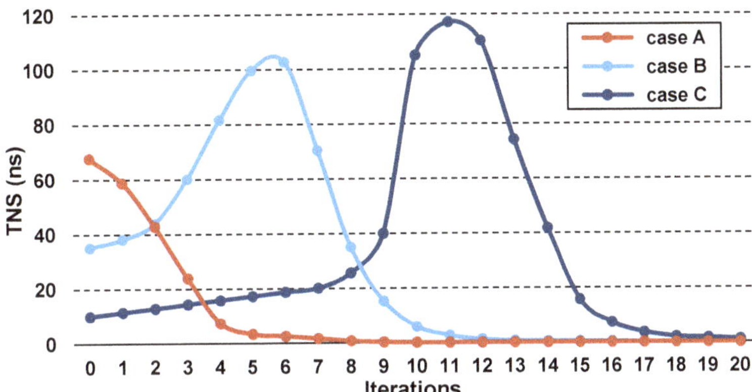

Figure 4. Evolution of TNS on each iteration of a LR-based gate sizer for three cases of the same design. In case A, design is not optimized. In cases B and C, it is partially optimized, thus exhibiting initially less TNS.

On the other hand, if the design had initially less TNS (case B), LR-based gate sizer prefers to improve power by degrading timing in the first six iterations before it starts solving timing violations and achieving timing closure in iteration eleven. Similarly, if LR-based gate sizing is applied on an already optimized design with only few timing violations (case C), it will first convert many nontiming critical paths to critical before actually reducing TNS to almost zero.

Regardless of the initial TNS, LR-based gate sizing is powerful enough to solve all timing violations. However, due to improper initialization, it fails to do this fast in cases that it should have. Therefore, for the case of partially optimized designs with a small set of timing violations, like case C of Figure 4, we need to derive an incremental version of the LR-based gate sizer that would achieve high quality-of-results and fast convergence.

3.2. What Can We Do about It?

To improve the applicability of the LR-based discrete gate sizer in an incremental optimization context, we propose an efficient method for initializing the values of the LMs so that the value of each LM is adaptive to the initial design state. In this way, the LMs are not set to an a priori chosen value but the values of the LMs would reflect the proper timing criticality of each gate relative to its already selected size, as seen near the end of the physical synthesis flow. The proposed approach is nonintrusive, since it deals only with the initialization of the LMs, and can be used with any LR-based gate sizer [15,31,32].

Determining the initial values of the LMs should not be based solely on the criticality of the corresponding timing arcs. Assume, for instance, that the design contains a very large gate that contributes a lot to its leakage power and currently has zero timing violations. In fact, we may assume that its output pin has a small positive slack. If we assign to this gate a small initial LM due to its positive slack, we would lead the optimizer to downsize it in the first iterations to save power. This choice may seem reasonable but it fails to answer one critical question: why this gate has not been downsized earlier by the multiple optimization steps that preceded? The most probable answer is that this gate originally belonged to a set of critical timing paths. Optimizing those paths in the first steps of the flow, resulted in selecting for this gate a fast (with small delay) but large cell. Thus, any trial to reduce its size at the end of the flow would directly translate to new timing violations.

Based on this intuition, we choose to initialize the LMs following a balanced approach. We assign increased LMs to timing arcs that are either critical at the moment or belong to high-power cells assuming that those cells may were timing critical in the past. This approach may lead to a temporary power overhead to cells that are indeed not critical but remained large for the wrong reasons (e.g., a previously applied optimization skipped them to save runtime). However, the first iterations of LR-based gate sizer would identify this by gradually reducing their corresponding LMs thus turning them to good candidates for power reduction.

The initial value for the LM of timing arc $i \rightarrow j$ is set to:

$$\lambda_{ij} = \left(\frac{a_i + d_{ij}}{a_j} \frac{P(g)}{\min P(g)} \right)^K \forall \text{ arc } i \rightarrow j \text{ of gate } g \tag{5}$$

Gate g refers to the gate where the timing arc $i \rightarrow j$ belongs. The starting value for each LM is the product of two ratios. The first ratio reveals the timing criticality of the arc $i \rightarrow j$. If the corresponding timing arc is responsible for the late arrival time at the output pin of gate j, the sum of a_i and the delay $d_{i,j}$ will be equal to a_j thus setting the ratio to 1. In any other case, a_j will be greater than the numerator and thus the ratio will result to a value less than 1 signifying the non criticality of the arc. The second ratio describes how much more power the current version of the cell spends $P(g)$ relative to the minimum possible leakage power that it can spend using any compatible library cell for gate g. Overall, when timing critical arcs are coupled with high power cells will get much greater LM values. The exponent K helps to make faster the assigned LMs' values, and we empirically set it to $K = 2$.

Similarly, for the LMs that correspond to the timing arcs $i \rightarrow k$, where k is a timing endpoint:

$$\lambda_{ik} = \left(\frac{a_k}{r_k} \frac{\Sigma_{gates} P(g)}{\Sigma_{gates} \min P(g)} \right)^K \forall \text{ timing endpoint } k \tag{6}$$

If the signal arrives at the timing endpoint k earlier than its required time r_k, i.e., $a_k < r_k$, signaling that there is no timing violation, the first ratio will result in a value less than one. On the contrary, in cases that late timing is violated, with $a_k > r_k$, the first ratio will be as big as the actual violation. For the power ratio in the case of timing endpoints, we suggest that it should consider the design as a whole. For this reason, the power ratio that is multiplied to the the timing ratio, divides the current total leakage power of the design relative to the minimum leakage power that the design can achieve after replacing each gate with a minimum leakage power cell. This ratio actually quantifies how far the design is from its virtually minimum leakage power.

Once the LMs were initialized, they need to be scaled to respect the KKT conditions as described in (3). Following (3) the sum of LMs of the output timing arcs of a gate should be equal to the timing arcs at its input. For instance, for the gate 6 shown in Figure 2, we should guarantee that $\lambda_{67} + \lambda_{68} = \lambda_{36} + \lambda_{46}$.

To achieve this, each one of the input LMs λ_{36} and λ_{46} receive a percentage of the sum of output LMs $\lambda_{67} + \lambda_{68}$. How much of the sum of output LMs would flow to each input LM is determined by the initial values λ_{36}^{init} and λ_{46}^{init} of timing arcs $3 \rightarrow 6$ and $4 \rightarrow 6$, respectively.

$$\lambda_{36} = \frac{\lambda_{36}^{init}}{\lambda_{36}^{init} + \lambda_{46}^{init}} (\lambda_{67} + \lambda_{68}) \qquad \lambda_{46} = \frac{\lambda_{46}^{init}}{\lambda_{36}^{init} + \lambda_{46}^{init}} (\lambda_{67} + \lambda_{68}) \tag{7}$$

The initial values of λ_{36}^{init} and λ_{46}^{init} are derived using Equation (5). When all gates were visited in reverse topological order and the LMs of the timing endpoints are propagated internally, the optimization can start.

4. Results

The proposed method was implemented in C++ inside the open-source RSyn physical design framework [44] after extending it for multicorner timing analysis. The evaluation involves already optimized benchmarks with only few timing violations. For this purpose, we used the fully optimized versions of the benchmarks of the ISPD 2013 gate sizing contest [37]. Those designs exhibit closed timing and minimized leakage power. To introduce additional timing violations, we randomly changed the resistance and capacitance of each net by $\pm10\%$, thus mimicking local rerouting operation at the end of the physical synthesis flow.

Our approach is experimentally validated using the benchmarks of the ISPD 2013 gate sizing contest considering a single and a multiple-corner scenario. For the case of multiple corners, we created two artificial (but realistic) timing libraries representing the fast (timing derate 1.05) and the slow version (timing derate 0.95) of the main typical library used in the single-corner case. Each cell in timing library has 10 sizes available at 3 Vth, with a total of 30 sizes per cell.

4.1. Quality-of-Results and Runtime Comparisons

Initially, we report the quality-of-results achieved for the proposed method (New) relative a state-of-the-art LR-gate sizer [31] (called Base) without allowing it to reset the state of the design. In other words, the optimization flow is the same as depicted in Figure 3 without performing the initial sizing step. Both cases actually utilize the same LR-based gate sizer. Their only difference is on how they initialize the value of the LMs. The obtained results are shown in Table 1 for single corner designs and in Table 2 for multicorner designs. Columns initially correspond to the design produced after randomly perturbing the resistance and capacitance of the wires. In all cases, the optimization stops if the improvement in terms of timing and leakage power across two iterations is less than 1%. Tables 1 and 2 report the late Worst Negative Slack (WNS), the late TNS and the total leakage power of each design under single and multiple corners, respectively. The final reported timing results are validated by OpenTimer [45]. ISPD2013 benchmarks do not exhibit early timing violations, and thus, early timing information is omitted.

The first noticeable result is that "New" offers better timing results than "Base" in the majority of the designs. With the proposed LM initialization, WNS is further decreased by 24% on average, while TNS is improved by more than 36% on average compared to the corresponding results of "Base" with only one corner. In multicorner designs, "New" helps improve WNS by a further 27% on average, while TNS improves by more 39%. In these cases, when timing slack reported is zero it means that timing constraints are satisfied in all corners. In all other cases, timing refers to the negative slack of the most critical corner.

"New" also achieves slightly better leakage power than "Base". For fair comparison, we take into account only the leakage power savings from designs where both the "Base" and the "New" flow succeeded to resolve all timing violations. In those cases, in single corner designs "New" is 2% better on average, and 1% better on average in multicorner designs. The reason for choosing only the timing closed designs is that whenever there are timing violations, the design's power is lower than the power of the design with closed timing.

Figure 5 compares the two approaches in terms of runtime when the designs have one corner. All experiments were performed on the same Linux-based workstation using a 3.6 GHz Intel Core i7-4790 with four cores and 32 GB of RAM. "New" is able to save up to 45% of runtime on average achieving also better quality-of-results. In terms of absolute runtime, the single corner "Base" finishes optimizing all designs in 9 h, while the proposed flow needs 5 h for the same task. The runtime of "Base" and "New" methods for designs usb_phy (slow and fast) is similar due to their small size of the designs.

Table 1. Timing and leakage power of all designs under single corner initially (Init) and at end of incremental LR-based sizer without (Base) and with (New) proposed LM initialization.

| Design | #Cells | Single Corner | | | | | | | | |
| | | Late WNS (ps) | | | Late TNS (ps) | | | Leakage (mW) | | |
		Init	Base	New	Init	Base	New	Init	Base	New
usb_phy_slow	623	−1.53	0.00	0.00	−1.53	0.00	0.00	1	1	1
usb_phy_fast		−0.61	0.00	0.00	−0.61	0.00	0.00	2	2	2
pci_bridge32_slow	30,763	−11.21	0.00	0.00	−333.10	0.00	0.00	58	58	58
pci_bridge32_fast		−16.66	−0.44	0.00	−614.66	−0.96	0.00	98	97	100
fft_slow	33,792	−16.35	0.00	0.00	−320.92	0.00	0.00	88	88	87
fft_fast		−18.18	−6.58	−1.88	−234.28	−63.37	−4.25	217	228	228
cordic_slow	42,937	−13.99	−14.43	−1.24	−801.84	−116.70	−2.11	306	349	309
cordic_fast		−13.26	−4.26	−6.94	−752.72	−30.00	−31.40	1139	1142	933
des_perf_slow	113,346	−30.40	−1.88	0.00	−11,920.00	−5.26	0.00	449	410	420
des_perf_fast		−25.80	−3.51	−4.10	−11,412.20	−49.94	−8.69	609	522	556
edit_dist_slow	129,227	−54.44	0.00	0.00	−21,881.50	0.00	0.00	452	447	445
edit_dist_fast		−63.59	−3.34	0.00	−36,639.50	−15.16	0.00	624	630	610
matrix_mult_slow	159,642	−44.00	0.00	0.00	−3292.93	0.00	0.00	481	487	476
matrix_mult_fast		−33.07	0.00	0.00	−2694.75	0.00	0.00	1056	1230	1020
netcard_slow	984,094	−30.19	0.00	0.00	−1477.58	0.00	0.00	5160	5101	5102
netcard_fast		−28.97	0.00	0.00	−6394.27	0.00	0.00	5203	5144	5141
Average		−25.14	−2.15	−0.89	−6173.27	−17.59	−2.90	996	996	968

Table 2. Timing and leakage power of all designs under multiple corners initially nd at end of incremental LR−based sizer without (Base) and with (New) proposed LM initialization.

| Design | Multiple Corners | | | | | | | | |
| | Late WNS (ps) | | | Late TNS (ps) | | | Leakage (mW) | | |
	Init	Base	New	Init	Base	New	Init	Base	New
usb_phy_slow	−0.03	0.00	0.00	−0.03	0.00	0.00	1	1	1
usb_phy_fast	−6.38	−4.99	0.00	−14.39	−8.57	0.00	3	2	3
pci_bridge32_slow	−14.76	0.00	0.00	−485.44	0.00	0.00	60	59	59
pci_bridge32_fast	−21.40	−4.25	0.00	−280.77	−14.78	0.00	194	151	153
fft_slow	−10.74	−0.14	0.00	−194.37	−0.27	0.00	96	97	98
fft_fast	−8.21	0.00	0.00	−449.16	0.00	0.00	356	426	391
cordic_slow	−24.57	−0.68	−2.06	−1000.51	−1.09	−2.06	518	561	527
cordic_fast	−122.26	−92.07	−66.50	−5412.47	−2954.33	−1710.28	2604	3189	3220
des_perf_slow	−34.07	−29.24	−14.08	−11,391.80	−42.13	−26.27	723	704	715
des_perf_fast	−77.49	−46.25	−33.07	−19,884.50	−737.60	−216.15	1272	926	1038
edit_dist_slow	−67.66	0.00	0.00	−36,892.70	0.00	0.00	477	473	471
edit_dist_fast	−68.96	−11.22	0.00	−39,745.10	−77.41	0.00	766	791	754
matrix_mult_slow	−43.79	0.00	0.00	−3254.51	0.00	0.00	576	591	574
matrix_mult_fast	−36.16	−45.23	−33.02	−3243.41	−107.50	−41.07	1876	2357	2302
netcard_slow	−42.25	0.00	0.00	−2251.09	0.00	0.00	5163	5105	5105
netcard_fast	−28.96	−1.23	0.00	−10,606.80	−2.34	0.00	5245	5187	5183
Average	−37.98	−14.71	−9.3	−8444.19	−246.63	−124.74	1246	1289	1287

Similarly, Figure 6, reveals the runtime savings of the proposed approach in a multi-corner timing scenario. The runtime of "New" is by 42% on average less than the average runtime of "Base". Multi-corner "Base" finishes optimizing all designs in 12 h. When the proposed initialization method is used, the total execution time is reduced to 6hrs. The overall increased execution time of multicorner optimization relative to the single corner scenario is due to the increased runtime of performing timing analysis on all corners.

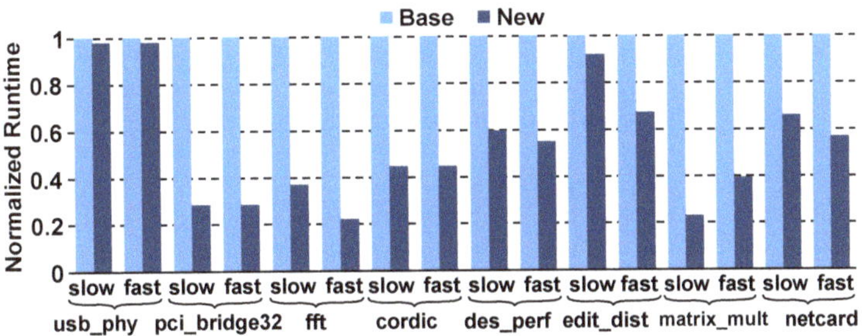

Figure 5. Runtime of both methods under comparison for all benchmarks when considering only one corner. Runtime is normalized to runtime of "Base" run. In all cases, "New" allows faster convergence saving up to 45% execution time on average in single corner benchmarks.

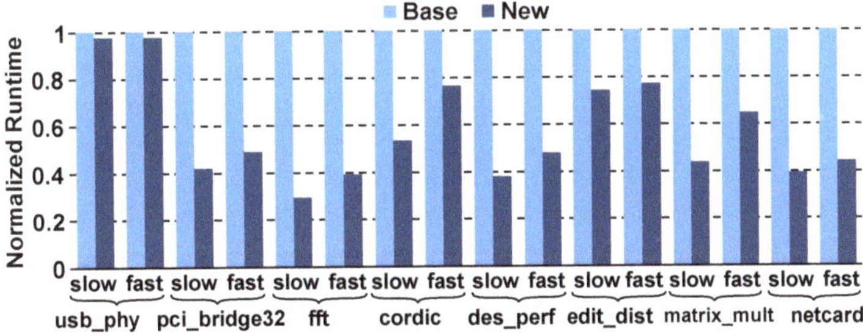

Figure 6. Runtime of both methods under comparison for all benchmarks when considering multiple corners. Runtime is normalized to runtime of "Base" run. In all cases, "New" allows faster convergence saving up to 42% on average with multiple corners.

4.2. Exploring in Depth the Proposed LM Initialization

Additional experimental results reveal that the way the LMs are initialized is crucial for the fast convergence and the overall timing QoR. Figure 7 compares the normalized late TNS of fft_fast design with one corner for different exponents K of the proposed Equations (5) and (6). As the value of exponent K increases, higher LMs are initialized to the timing critical arcs of the design. This means that the timing improves faster with better overall QoR. For all our experiments, we selected $K = 2$ because exponent values above $K = 2$ does not improve any further the QoR.

To observe more clearly how the proposed LM initialization helps the convergence of an LR-based gate sizer, we monitor the evolution of TNS across consecutive iterations initializing the LMs to different values. For the des_perf_fast design, shown in Figure 8, "Base" starts degrading the timing until iteration four where the TNS reaches 80 ns. From this point, the actual optimization starts and the timing closure is achieved in iteration ten. Applying the proposed Equations (5) and (6) ("New"), the optimizer starts reducing the timing violations immediately without degrading the initial state of the design and the timing constraints are met in iteration five. To further evaluate our work, we also tried to initialize the LMs to different values where the starting value of each LM was randomly selected ("Random"). In this case, the peak of the TNS is increased compared to the "Base" run. More specifically, the TNS in iteration four is increased from 80 ns to 110 ns, and thus, 3 more iterations were needed, compared to that of "Base", to close the timing. Finally, we tested the performance of the LR-based gate sizer adopting the initialization method of work [31], in which the authors start all the LMs from 12. Even though this modification could slightly decrease the highest value of the TNS (compared to "Base"), the optimization showed slower convergence. The TNS improvement delayed to start and the timing constraints were finally met after multiple iterations, in iteration 15. From all the LM initialization trials, "New" showed the fastest convergence of all closing the timing really soon. Similar results are obtained for all other designs. The proposed LM initialization successfully "predicts" the value of the LM that fits better to the status of the design, thus avoiding unnecessary power reductions at the first iterations that would hurt timing initially and delay convergence later on.

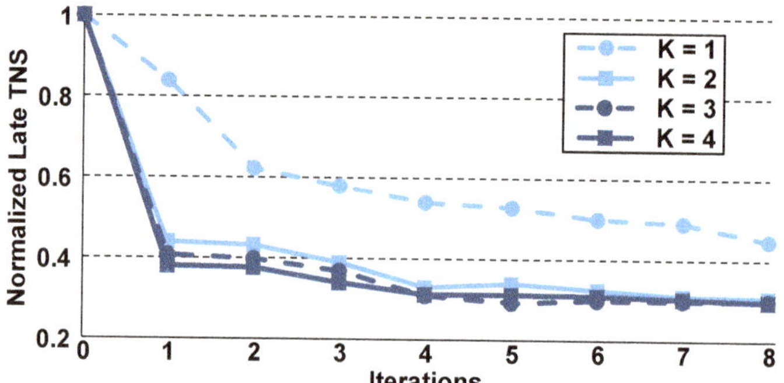

Figure 7. TNS comparison for different values of exponent *K* for LM initialization as proposed in Equations (5) and (6) for representative design, fft_fast. Higher values of *K* increase LMs of critical arcs leading to faster TNS improvement during optimization iterations. Beyond *K* = 2, there are not sufficient savings.

To be certain for the quality-of-results of the proposed approach, we repeated the same experiment for each benchmark 100 times. Each time, the methods under comparison were applied on designs produced after perturbing randomly the wire parasitics of the already optimized version of each benchmark. The histogram of TNS for the initial design, and the ones produced after applying "Base" and "New" methods are depicted in Figure 9 for benchmark fft_fast, while similar results are obtained for all other benchmarks.

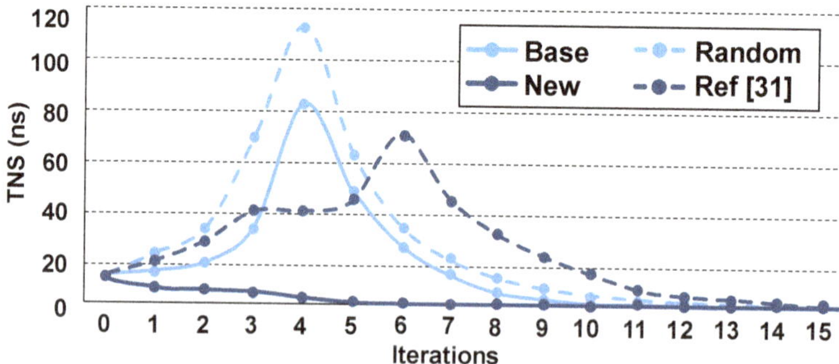

Figure 8. Progression of late TNS on des_perf_fast design using different LM initialization methods; initializing LMs to 1 (Base), using proposed method (New), initializing each LM to a random value (Random) and using initialization of [31] (Ref [31]).

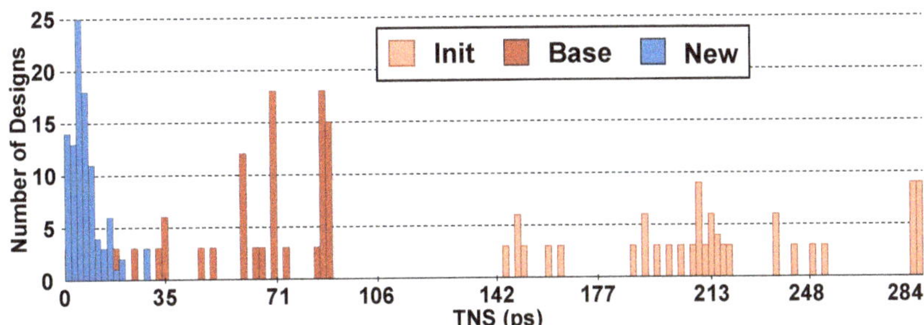

Figure 9. Histogram of late TNS initially (Init) and at end of LR-based gate sizing without (Base) and with (New) proposed LM initialization. Histograms correspond to 100 versions of fft_fast with randomly perturbed RC characteristics.

TNS histograms reveal that both approaches successfully decreased the original TNS. "Base" decreased the mean of initial TNS from 225 ps to 65 ps, while "New" managed to compress the TNS histogram to the left side of the diagram, with the majority of samples gathered close to 5 ps.

4.3. Optimization with a Restricted Number of Available Gate Sizes

For completeness, we evaluated both "Base" and "New" methods under comparison in a more restrictive scenario. In this case, gate sizing is only allowed to resize cells only to their next bigger or smaller size without limiting V_T swapping options, since they do not alter the physical layout. This restriction makes sense at the final steps of physical design flow to preserve as much as possible the already defined detailed wire routes. The obtained results of single corner and multicorner benchmarks are depicted in Table 3 and in Table 4, respectively. Besides the restricted availability of gate sizes, "New" achieves considerable improvements. In single corner designs, late WNS is improved by 36% on average while the savings in TNS reach 39% on average, when compared to that of the baseline single corner LR-based gate sizer. In terms of leakage power, the restricted "New" method achieves less leakage power by 2% on average, when considering only the designs without negative slack at both methods under comparison. For multiple corners, late WNS is improved by 35% on average, while late TNS improves by 39% on average when compared to the corresponding timing results of "Base". Also "New" achieves slightly less leakage power by 2% on average.

Table 3. Timing and leakage power of all designs under single corner with gate size selection restriction without (Base) and with (New) proposed LM initialization.

	Single Corner					
Design	**Late WNS (ps)**		**Late TNS (ps)**		**Leakage (mW)**	
	Base	**New**	**Base**	**New**	**Base**	**New**
usb_phy_slow	0.00	0.00	0.00	0.00	1	1
usb_phy_fast	0.00	0.00	0.00	0.00	2	2
pci_bridge32_slow	0.00	0.00	0.00	0.00	58	58
pci_bridge32_fast	−1.65	0.00	−6.13	0.00	98	98
fft_slow	0.00	0.00	0.00	0.00	88	87
fft_fast	−6.87	−1.01	−20.18	−2.24	224	221
cordic_slow	−8.79	−2.96	−67.21	−2.96	378	310
cordic_fast	−17.06	−2.73	−133.10	−4.81	1209	942
des_perf_slow	−27.50	−1.40	−67.53	−4.52	480	464
des_perf_fast	−14.41	−7.61	−47.42	−23.30	637	611

Table 3. *Cont.*

| Design | Single Corner | | | | | |
| | Late WNS (ps) | | Late TNS (ps) | | Leakage (mW) | |
	Base	New	Base	New	Base	New
edit_dist_slow	0.00	0.00	0.00	0.00	450	449
edit_dist_fast	−20.77	−1.95	−698.80	−2.16	623	619
matrix_mult_slow	0.00	0.00	0.00	0.00	478	479
matrix_mult_fast	0.00	0.00	0.00	0.00	1174	1020
netcard_slow	0.00	0.00	0.00	0.00	5152	5153
netcard_fast	0.00	0.00	0.00	0.00	5197	5194
Average	−6.07	−1.10	−65.02	−2.50	1016	982

Table 4. Timing and leakage power of all designs under multiple corners with gate size selection restriction without (Base) and with (New) proposed LM initialization.

| Design | Multiple Corners | | | | | |
| | Late WNS (ps) | | Late TNS (ps) | | Leakage (mW) | |
	Base	New	Base	New	Base	New
usb_phy_slow	0.00	0.00	0.00	0.00	1	1
usb_phy_fast	−12.81	0.00	−42.00	0.00	2	2
pci_bridge32_slow	0.00	0.00	0.00	0.00	62	60
pci_bridge32_fast	−22.20	−21.24	−189.58	−154.97	170	170
fft_slow	0.00	0.00	0.00	0.00	100	98
fft_fast	−22.48	0.00	−92.88	0.00	366	365
cordic_slow	−1.48	0.00	−1.86	0.00	705	516
cordic_fast	−113.97	−112.70	−5604.24	−4867.80	3325	3389
des_perf_slow	−30.88	−18.45	−207.30	−125.34	728	713
des_perf_fast	−68.24	−47.37	−1520.11	−386.81	1205	1229
edit_dist_slow	0.00	0.00	0.00	0.00	478	477
edit_dist_fast	−3.11	−0.48	−3.11	−0.85	824	758
matrix_mult_slow	0.00	0.00	0.00	0.00	602	580
matrix_mult_fast	−26.23	−27.31	−42.98	−43.54	2214	2154
netcard_slow	0.00	0.00	0.00	0.00	5172	5158
netcard_fast	−4.70	0.00	−7.60	0.00	5250	5236
Average	−19.13	−14.22	−481.98	−348.71	1325	1307

5. Conclusions

Efficient incremental and minimally disruptive optimization steps at the end of the design flow are crucial for the overall success of automated physical synthesis. In this work, instead of relying on custom-made timing and power optimization heuristics, we leverage, for the first time—to the best of our knowledge—LR-based optimizers used for the global optimization of the design as fast incremental optimizers after appropriate initialization. Initialization involves selecting appropriate values for the LMs after taking into account both their timing criticality, in a multicorner context, as well as the current size of the gates. In this way, we expedite successfully the convergence of the LR-based gate sizer, when applied in an incremental optimization context, without affecting any part of its internal functions and without reducing the achieved quality-of-results.

Experimental results also showed that relying on constant LM initialization values as done by similar state-of-the-art optimizers or using randomly selected constants do not achieve the smooth convergence needed in the case of last-mile incremental timing optimizations. Initializing the LMs with hand-selected constants provides an inaccurate picture of the design to the LR optimizer. This picture translates to unnecessary power reductions and timing degradation at the beginning of the optimization and inevitably leads to many more iterations before reconverging back to an timing optimized solution. This deficit was corrected by the proposed approach and allows LR-based global optimizers to be successfully used as fast incremental timing optimizers.

Our future plans are to incorporate the proposed LM initialization strategy into similar timing-driven placement engines that tradeoff placement density and wirelength for better timing performance.

Author Contributions: All authors have contributed equally and substantially to all parts of this manuscript. Conceptualization, D.M. and G.D.; methodology, D.M. and G.D.; writing—original draft preparation, D.M. and G.D. All authors have read and agreed to the published version of the manuscript.

Funding: Dimitrios Mangiras is supported by the Onassis Foundation—Scholarship ID: G ZO 014-1/2018-2019.

Institutional Review Board Statement: Not applicable.

Informed Consent Statement: Not applicable.

Data Availability Statement: Not applicable.

Conflicts of Interest: The authors declare no conflict of interest.

References

1. Lavagno, L.; Martin, G.; Markov, I.L.; Scheffer, L.K. *Electronic Design Automation for IC Implementation, Circuit Design, and Process Technology*; Taylor and Francis Group: Boca Raton, FL, USA, 2016.
2. Liu, Y.; Hu, J.; Shi, W. Multi-Scenario Buffer Insertion in Multi-Core Processor Designs. In Proceedings of the 2008 International Symposium on Physical Design, Portland, OR, USA, 13–16 April 2008; pp. 15–22.
3. Roy, S.; Liu, D.; Um, J.; Pan, D.Z. OSFA: A new paradigm of gate-sizing for power/performance optimizations under multiple operating conditions. In Proceedings of the Design Automation Conference (DAC), San Francisco, CA, USA, 8–12 June 2015; pp. 1–6.
4. MacDonald, N.D. Timing Closure in Deep Submicron Designs. In Proceedings of the Design Automation Conference (DAC), Anaheim, CA, USA, 13–18 July 2010.
5. Chinnery, D.G.; Keutzer, K. Linear Programming for Sizing, Vth and Vdd Assignment. In Proceedings of the International Symposium on Low Power Electronics and Design (ISLPED), San Diego, CA, USA, 8–10 August 2005; pp. 149–154.
6. Spindler, P.; Schlichtmann, U.; Johannes, F.M. Abacus: Fast legalization of standard cell circuits with minimal movement. In Proceedings of the International Symposium on Physical Design (ISPD), Portland, OR, USA, 13–16 April 2008; pp. 47–53.
7. Puget, J.C.; Flach, G.; Reis, R.; Johann, M. Jezz: An effective legalization algorithm for minimum displacement. In Proceedings of the Symposium on Integrated Circuits and Systems Design (SBCCI), Salvador, Brazil, 31 August–4 September 2015; pp. 1–5.
8. Chowdhary, A.; Rajagopal, K.; Venkatesan, S.; Cao, T.; Tiourin, V.; Parasuram, Y.; Halpin, B. How Accurately Can We Model Timing in a Placement Engine? In Proceedings of the ACM/IEEE Design Automation Conference (DAC), Anaheim, CA, USA, 13–17 June 2005; pp. 801–806.
9. Alpert, C.; Chu, C.; Gandham, G.; Hrkić, M.; Hu, J.; Kashyap, C.; Quay, S. Simultaneous Driver Sizing and Buffer Insertion Using a Delay Penalty Estimation Technique. In Proceedings of the International Symposium on Physical Design (ISPD), San Diego, CA, USA, 7–10 April 2002; pp. 104–109.
10. Stefanidis, A.; Mangiras, D.; Nicopoulos, C.; Chinnery, D.; Dimitrakopoulos, G. Autonomous Application of Netlist Transformations inside Lagrangian Relaxation-based Optimization. *IEEE Trans. Comput.-Aided Des. Integr. Circuits Syst.* **2021**, *40*, 1672–1686. [CrossRef]
11. Jiang, Y.; Sapatnekar, S.S.; Bamji, C.; Kim, J. Interleaving buffer insertion and transistor sizing into a single optimization. *IEEE Trans. VLSI Syst.* **1998**, *6*, 625–633. [CrossRef]
12. Fishburn, J.P. Clock Skew Optimization. *IEEE Trans. Comput.* **1990**, *39*, 945–951. [CrossRef]
13. Kim, S.; Do, S.; Kang, S. Fast Predictive Useful Skew Methodology for Timing-Driven Placement Optimization. In Proceedings of the ACM/IEEE Design Automation Conference (DAC), Austin, TX, USA, 18–22 June 2017; pp. 55:1–55:6.
14. Fishburn, J.P.; Dunlop, A.E. TILOS: A posynomial programming approach to transistor sizing. In *ICCAD 2003*; Springer: Boston, MA, USA, 2003.
15. Ozdal, M.M.; Burns, S.; Hu, J. Algorithms for Gate Sizing and Device Parameter Selection for High-Performance Designs. *IEEE Trans. CAD* **2012**, *31*, 1558–1571. [CrossRef]
16. Coudert, O. Gate Sizing for Constrained Delay/Power/Area Optimization. *IEEE Trans. VLSI Syst.* **1997**, *5*, 465–472. [CrossRef]

17. Nguyen, D.; Davare, A.; Orshansky, M.; Chinnery, D.; Thompson, B.; Keutzer, K. Minimization of Dynamic and Static Power Through Joint Assignment of Threshold Voltages and Sizing Optimization. In Proceedings of the 2003 International Symposium on Low Power Electronics and Design (ISLPED '03), Seoul, Korea, 25–27 August 2003; pp. 158–163.

18. Bhattacharya, K.; Ranganathan, N. A Linear Programming Formulation for Security-Aware Gate Sizing. In Proceedings of the ACM Great Lakes Symposium on VLSI (GLSVLSI '08), Orlando, FL, USA, 4–6 May 2008; pp. 273–278.

19. Berkelaar, M.; Jess, J. Gate sizing in MOS digital circuits with linear programming. In Proceedings of the European Design Automation Conference, Glasgow, UK, 12–15 March 1990; pp. 217–221.

20. Jeong, K.; Kahng, A.B.; Yao, H. Revisiting the linear programming framework for leakage power vs. performance optimization. In Proceedings of the 2009 10th International Symposium on Quality Electronic Design, San Jose, CA, USA, 16–18 March 2009; pp. 127–134.

21. Reimann, T.; Posser, G.; Flach, G.; Johann, M.; Reis, R. Simultaneous gate sizing and Vt assignment using Fanin/Fanout ratio and Simulated Annealing. In Proceedings of the 2013 IEEE International Symposium on Circuits and Systems (ISCAS), Beijing, China, 19–23 May 2013; pp. 2549–2552.

22. Daboul, S.; Hähnle, N.; Held, S.; Schorr, U. Provably Fast and Near-Optimum Gate Sizing. *IEEE Trans. Comput.-Aided Des. Integr. Circuits Syst.* **2018**, *37*, 3163–3176. [CrossRef]

23. Hu, S.; Ketkar, M.; Hu, J. Gate Sizing For Cell Library-Based Designs. In Proceedings of the 2007 44th ACM/IEEE Design Automation Conference, San Diego, CA, USA, 4–8 June 2007; pp. 847–852.

24. Ozdal, M.M.; Burns, S.; Hu, J. Gate sizing and device technology selection algorithms for high-performance industrial designs. In Proceedings of the 2011 IEEE/ACM International Conference on Computer-Aided Design (ICCAD), San Jose, CA, USA, 7–10 November 2011; pp. 724–731.

25. Rahman, M.; Tennakoon, H.; Sechen, C. Library-Based Cell-Size Selection Using Extended Logical Effort. *IEEE Trans. Comput.-Aided Des. Integr. Circuits Syst.* **2013**, *32*, 1086–1099. [CrossRef]

26. Liu, Y.; Hu, J. A New Algorithm for Simultaneous Gate Sizing and Threshold Voltage Assignment. *IEEE Trans. Comput.-Aided Des. Integr. Circuits Syst.* **2010**, *29*, 223–234. [CrossRef]

27. Hu, J.; Kahng, A.B.; Kang, S.; Kim, M.C.; Markov, I.L. Sensitivity-guided metaheuristics for accurate discrete gate sizing. In Proceedings of the IEEE International Conference CAD, San Jose, CA, USA, 5–8 November 2012; pp. 233–239.

28. Kahng, A.B.; Kang, S.; Lee, H.; Markov, I.L.; Thapar, P. High-performance Gate Sizing with a Signoff Timer. In Proceedings of the International Conference on Computer-Aided Design (ICCAD), San Jose, CA, USA, 18–21 November 2013; pp. 450–457.

29. Fatemi, H.; Kahng, A.B.; Lee, H.; Li, J.; Pineda de Gyvez, J. Enhancing sensitivity-based power reduction for an industry IC design context. *Integration* **2019**, *66*, 96–111. [CrossRef]

30. Lu, Y.C.; Nath, S.; Khandelwal, V.; Lim, S.K. RL-Sizer: VLSI Gate Sizing for Timing Optimization using Deep Reinforcement Learning. In Proceedings of the 2021 58th ACM/IEEE Design Automation Conference (DAC), San Francisco, CA, USA, 5–9 December 2021; pp. 733–738.

31. Flach, G.; Reimann, T.; Posser, G.; Johann, M.; Reis, R. Effective Method for Simultaneous Gate Sizing and Vth Assignment Using Lagrangian Relaxation. *IEEE Trans. Comput.-Aided Des. Integr. Circuits Syst.* **2014**, *33*, 546–557. [CrossRef]

32. Sharma, A.; Chinnery, D.; Bhardwaj, S.; Chu, C. Fast Lagrangian Relaxation Based Gate Sizing Using Multi-Threading. In Proceedings of the IEEE Inter. Conf. on Computer-Aided Design, Austin, TX, USA, 2–6 November 2015; pp. 426–433.

33. Sharma, A.; Chinnery, D.; Dhamdhere, S.; Chu, C. Rapid gate sizing with fewer iterations of Lagrangian Relaxation. In Proceedings of the 2017 IEEE/ACM International Conference on Computer-Aided Design (ICCAD), Irvine, CA, USA, 13–16 November 2017; pp. 337–343.

34. Livramento, V.S.; Guth, C.; Güntzel, J.L.; Johann, M.O. A Hybrid Technique for Discrete Gate Sizing Based on Lagrangian Relaxation. *ACM Trans. Des. Autom. Electron. Syst.* **2014**, *19*. [CrossRef]

35. Shklover, G.; Emanuel, B. Simultaneous Clock and Data Gate Sizing Algorithm with Common Global Objective. In Proceedings of the 2012 ACM International Symposium on International Symposium on Physical Design, Napa, CA, USA, 25–28 March 2012; pp. 145–152.

36. Li, L.; Kang, P.; Lu, Y.; Zhou, H. An efficient algorithm for library-based cell-type selection in high-performance. In Proceedings of the 2012 IEEE/ACM International Conference on Computer-Aided Design (ICCAD), San Jose, CA, USA, 5–8 November 2012; pp. 226–232.

37. Ozdal, M.; Amin, C.; Ayupov, A.; Burns, S.M.; Wilke, G.R.; Zhuo, C. An Improved Benchmark Suite for the ISPD-2013 Discrete Cell Sizing Contest. In Proceedings of the International Symposium on Physical Design, Stateline, NV, USA, 24–27 March 2013; pp. 168–170.

38. Mangiras, D.; Dimitrakopoulos, G. Incremental Lagrangian Relaxation based Discrete Gate Sizing and Threshold Voltage Assignment. In Proceedings of the 2021 10th International Conference on Modern Circuits and Systems Technologies (MOCAST), Thessaloniki, Greece, 5–7 July 2021; pp. 1–5.

39. Bhasker, J.; Chadha, R. *Static Timing Analysis for Nanometer Designs: A Practical Approach*; Springer: Boston, MA, USA, 2009.

40. Mangiras, D.; Stefanidis, A.; Seitanidis, I.; Nicopoulos, C.; Dimitrakopoulos, G. Timing-Driven Placement Optimization Facilitated by Timing-Compatibility Flip-Flop Clustering. *IEEE Trans. Comput.-Aided Des. Integr. Circuits Syst.* **2020**, *39*, 2835–2848. [CrossRef]

41. Berkelaar, M.; Buurman, P.; Jess, J. Computing the entire active area/power consumption versus delay tradeoff curve for gate sizing with a piecewise linear simulator. *IEEE Trans. Comput.-Aided Des. Integr. Circuits Syst.* **1996**, *15*, 1424–1434. [CrossRef]

42. Montiel-Nelson, J.; Sosa, J.; Navarro, H.; Sarmiento, R.; Núñez, A. Efficient method to obtain the entire active area against circuit delay time trade-off curve in gate sizing. *IEE Proc.-Circuits Dev. Syst.* **2005**, *152*, 133–145. [CrossRef]
43. Tennakoon, H.; Sechen, C. Nonconvex Gate Delay Modeling and Delay Optimization. *IEEE Trans. Comput.-Aided Des. Integr. Circuits Syst.* **2008**, *27*, 1583–1594. [CrossRef]
44. Flach, G.; Fogaça, M.; Monteiro, J.; Johann, M.; Reis, R. Rsyn: An Extensible Physical Synthesis Framework. In Proceedings of the International Symposium on Physical Design, Portland, OR, USA, 19–22 March 2017; pp. 33–40.
45. Huang, T.W.; Wong, M.D.F. OpenTimer: A high-performance timing analysis tool. In Proceedings of the IEEE/ACM International Conference on Computer-Aided Design (ICCAD), Austin, TX, USA, 2–6 November 2015; pp. 895–902.

MDPI
St. Alban-Anlage 66
4052 Basel
Switzerland
Tel. +41 61 683 77 34
Fax +41 61 302 89 18
www.mdpi.com

Technologies Editorial Office
E-mail: technologies@mdpi.com
www.mdpi.com/journal/technologies